in-Training

Stories from Tomorrow's Physicians

Volume 2

in-Training

Stories from Tomorrow's Physicians

Volume 2

Peer-edited narratives written by medical students
chronicling the major milestones of medical school

edited by

Ria Pal & Andrew Kadlec

Volume 2, 2018

PAGER PUBLICATIONS, INC.
a 501c3 non-profit literary corporation

in-Training: Stories from Tomorrow's Physicians, Volume 2

Published by Pager Publications, Inc. at pagerpublications.org.

Printed in the United States of America.

Cover art by Nita Chen.
Book design by Ajay Major.

Volume 2
First Printing: 2018

ISBN-13: 978-0-578-41080-7

To all those who have been so patient with us,
knowing that some generosity cannot be repaid.

Contents

Volume 2, 2018

in-Training Mission Statement

in-Training is the online peer-reviewed publication
for medical students,
founded in April 2012 by Ajay Major and Aleena Paul,
medical students at Albany Medical College.

in-Training is the agora of the medical student community,
the intellectual center for news, commentary, and the
free expression of the medical student voice.

in-Training seeks to:

enrich the medical education experience through self-reflection;

foster discourse among medical students;

and cultivate collaborative relationships between medical students
on the global stage.

Pager Publications, Inc. Mission Statement

Pager Publications, Inc. is a 501c3 nonprofit literary organization
that curates and supports peer-edited publications
for the medical education community.

The organization strives to provide students and educators
with dedicated spaces for the free expression of their distinctive voices.

Pager Publications, Inc. was officially incorporated in January 2015
by its founders Ajay Major, Aleena Paul and Erica Fugger
to provide administrative and financial support for
in-Training and other publications.

in-Training

the agora of the medical student community

Ria Pal and Andrew Kadlec, PhD
Editors-in-Chief

Ajay Major, MD, MBA and Aleena Paul, MD, MBA
Founders and Editors-in-Chief Emeriti

Lexy Adams and Ashten Duncan
Managing Editors

Editors

Samia Nawaz	Yuli Zhu	Daniel Fuchs
Tim Beck	Lytani Wilson	Brenna Brown
Visvarath Varadarajan	Amy Schettino	Michael Kung
Anup Bhattacharya	Anna Qian	Jessica Downing
Mark Shafarenko	Jason Bunn	Maddison Bibby
Jacob Kammerman	James Lee	Heba Albasha
Slavena Slave Nissan	Ileana Horattas	Sandy Tadros
Kshama Bhyravabhotla	Brent Schnipke	Melanie Watt
Omar Ghabra	Angelica D'Aiello	Elaine Hsiang
Sarah Garvey	Steph Cockrill	Olivia Abbate

Nisha Pradhan
Chief Social Media Manager

Lauren Bojarski and Clara Thomson
Social Media Managers

Editors Emeriti

Phyllis Ying	Luke Fraley	Dragos Rezeanu
Samantha Margulies	Romela Petrosyan	Nikki Nametz
Nina Nguyen	Evan Torline	Emily Lu
Lindsey McDaniel	Roshini Selladurai	Brent Bjornsen
Claire Drom	Lisa Tran	Theresa Yang
Anne Nzuki	Daniel Coleman	Amol Utrankar
Chris Deans	David Yu	Mansi Sheth
Diane Brackett	Hormuz Nicolwala	Eric Donahue
Natalie Wilcox	Sanjay Salgado	Allison Lyle
Chelcie Soroka	Jimmy Tam Huy Pham	Steven Lange
Kate Joyce	Kimberly Ku	Nita Chen
Jarna Shah	Sasha Yakhkind	Nisha Hariharan
Will Jaffee	Laura Mucenski	Tolulope Omojokun
Matthew Lenardis	Francis Dailey	Dustin Nowotny

Acknowledgements

Medical school challenges us with how much we can carry,
and instructs us in how to share the load.

We are profoundly grateful to have had a team that excelled at both.

We thank our authors and artists,
who chose to share their work with *in-Training*,
and our editorial team who carefully brought it all into its final form.

Furthermore, this book would not be possible without
Ashten Duncan's diligence and enthusiasm;
Lexy Adams' impeccable management;
and Ajay Major's encouragement, experience,
and presumably caffeine-fueled hours of work.

We wish for each of them a story that is rich, well-told,
and only just begun.

We would also like to thank Aviva Hope Rutkin
for her copyediting prowess,
as well as Nita Chen for her beautiful cover art.

Preface

by Ria Pal and Andrew Kadlec, PhD

Editors of *in-Training: Stories from Tomorrow's Physicians, Vol. 2*

We are excited for you to read the reflections of medical students to the unique and transformative experience of medical school.

The *in-Training* team previously published a collection of stories from medical students across the globe, arranged according to common themes that regularly appear on the *in-Training* website. Those living the medical school experience directly, including students and educators, were the target audience of the first book, with the goal of fostering regular and open discussion about major milestones in medical education.

With this second publication from the *in-Training* team, we turn our attention to you: the friends, family, and loved ones of medical students. We know that it may be difficult to understand what the medical student in your life is experiencing, especially since their daily activities are constantly changing and riddled with unfamiliar terms. You know anatomy lab is a life-changing experience, but how does it actually affect students? And what about Step 1? You hear a lot (probably too much) about it but can't fully understand the rigor involved with preparing for that exam. And third year? Medical students can disappear for weeks at a time and may have to miss that important family event, and you're left wondering what exactly is happening during that time. It is our hope that this collection will provide you with a greater understanding of these major milestones to help you support and form a deeper connection to the medical student in your life.

We have arranged this collection in chronological fashion, starting with milestones encountered early on in medical school and concluding with stories that describe the overall medical school journey and were written at the closing hours of fourth year. Our editors selected those works that are representative, and hopefully informative, of each milestone.

This book presents a unique experience for each reader. We suggest turning to the section that is relevant to the issues that the medical student in your life may be currently facing. It is our hope that this collection will make our authors' stories accessible to many more readers, and we are grateful for the opportunity to share them here.

Yours-in-training,

Ria Pal and Andrew Kadlec, PhD
Editors-in-Chief Emeriti of *in-Training*
Class of 2018 at University of Rochester School of Medicine & Dentistry
and Class of 2020 at Medical College of Wisconsin

For many medical students, the first step inside the anatomy laboratory is transformative. There is something ineffable about the vulnerability and quiet reflection that many students experience when they are tasked with learning about the human body from those who came before them. The first incision is often regarded as a sacred moment, one in which you find yourself filled with a strange combination of trepidation and awe. Though the task of learning human anatomy is arduous, many students find reassurance in reflecting on the kind of lives their donors led. This reassurance is what makes the learning process so rewarding and drives home how great of an honor it is to learn in such a manner. By the end of the anatomy course, most students have a newfound appreciation for the natural order of things: life and death; living and dying; and being mortal.

Anatomy Lab

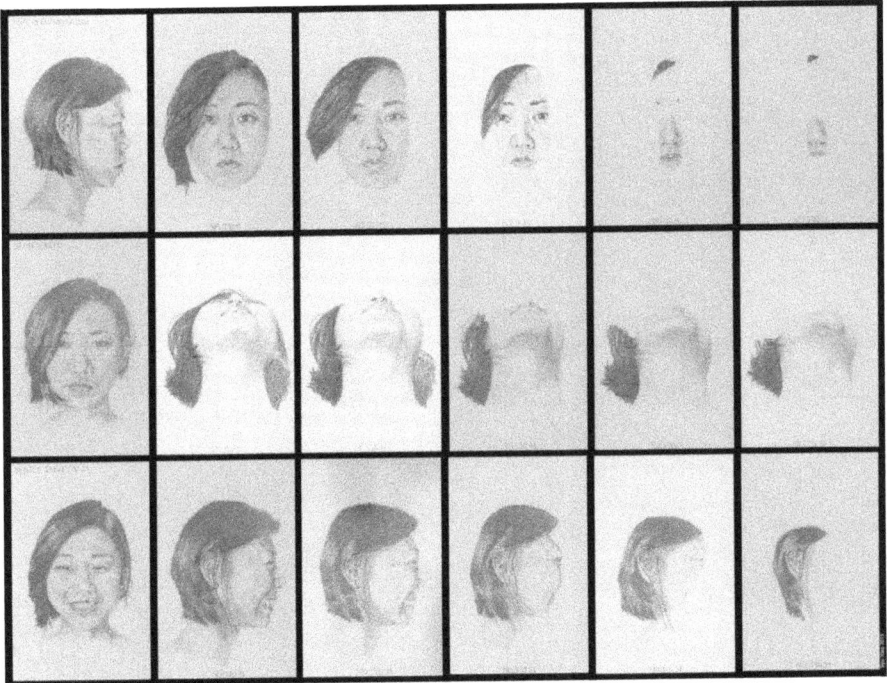

Atlas by Siyu Xiao
Class of 2018 at Yale School of Medicine

Anatomy Lab

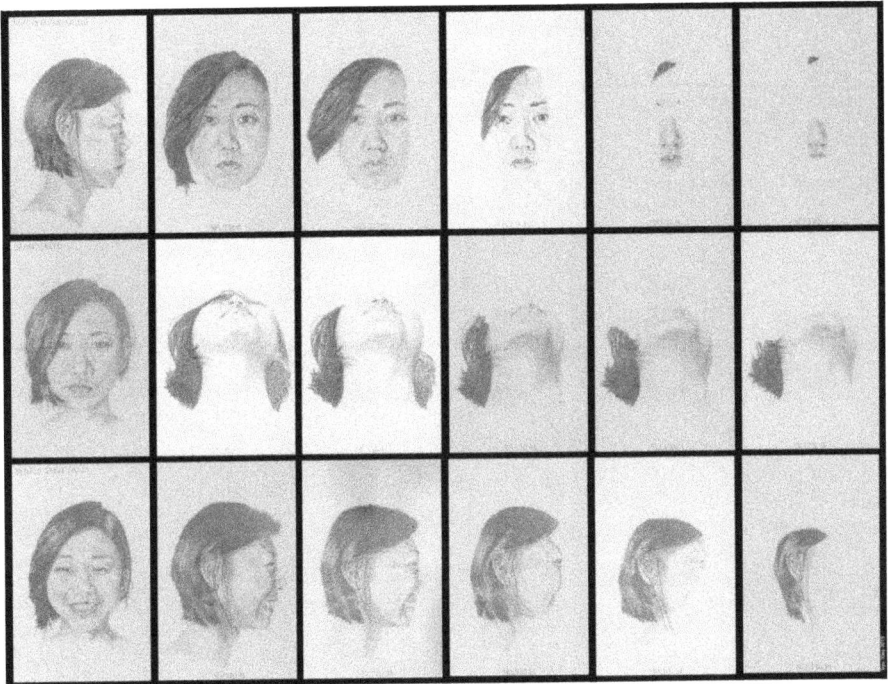

Atlas by Siyu Xiao
Class of 2018 at Yale School of Medicine

Our Cadavers, Ourselves

January 26, 2016

Leigh Finnegan
Perelman School of Medicine at the University of Pennsylvania
Class of 2019

M Y CADAVER HAS PINK fingernails. I saw them on the first day of class, after we pulled back the white plastic sheet with the number "22" scrawled on it with permanent marker and cut away the damp cloth that had been covering her cold skin. Her arms were folded across her chest, and on her fingers was a sparkly, ballet-pink polish, not chipped or peeling despite having been there for the 13 months since she'd died. I don't know why it's there. I don't know if she painted them thinking she was going to survive to enjoy it, or if she was someone who always wanted to look her best, even in death.

Thus was my introduction to gross anatomy, that rite of passage that separates medical students and physicians from those who have never had a look at what goes on under the skin. It's the class in which we take apart the human body to figure out how it works, not unlike the time when, as kids, my brother and I cut the fur off of our Furby toy to expose the underlying wires and plastic. I'd dreaded the class ever since interview season, when, clad in my suit and completely inappropriate footwear, I was thrust into a foul-smelling anatomy lab with my fellow interviewees, to see what it was like to be a real live medical student. The shock and revulsion almost made me question my entire decision to become a doctor.

In many ways, my dread was warranted. I never quite "got used to" anatomy the way older students told me I would, and I certainly never forgot that the person I was dissecting was just that — a person, formerly alive, whose name and personality I know nothing about, despite knowing exactly what her left renal vein looks like as it crosses her aorta on its way to her inferior vena cava. I've grown strangely attached to her over the last few months, which might be natural given the amount of time I've spent with her. She was a larger woman, but every time a professor or fellow student makes a comment about her weight, I find myself getting offended and angry, as though someone has judged and insulted a dear friend of mine without bothering to learn anything about her.

For the first month, I didn't know what her face looked like. Out of respect for cadavers and students alike, her head was covered with cloth until we began dissection of the head and neck. When we pulled it back, she looked peaceful, calm and undeniably human, despite the grayish undertone of skin lacking blood to perfuse it. That's the face I try to keep in my memory, even though it didn't look that way for long; within a few weeks, it had been skinned, excavated, and, eventually, cut entirely in half, hacked through by my dissection partners and me with a rusty saw.

The only thing I know for certain about my cadaver's life, when her heart still sent blood coursing through her arteries and ions still electrified her nerves, is that she suffered. On the first day of class, I saw the hernia in her abdomen, her viscera pulling through it like a baby on the verge of being born, likely a complication from the surgery that left a scar running down the length of her abdomen. I saw the way the ovarian cancer that killed her had wreaked havoc on her organs, leaving her intestines thickened and stuck to one another, her liver so teeming with metastases that when we tried to lift it up it all but disintegrated. I hope she had family around, that they comforted her when she felt weak, that she maintained an attitude as bright and sparkly as the polish on her fingernails. I hope, but I'll never know.

Despite taking up just seven or so hours per week, I find anatomy lab leaking into the rest of my life. I dream about it frequently, sometimes about my cadaver, up and walking around like a living human, but with the leathery skin of a preserved body and a stiff, robot-like gait reminiscent of the rigor mortis I've always known her to have. During our unit on limbs, I dreamt about cutting off my fingers to donate to somebody else who needed them, and woke up frantically flexing and extending the digits of my left hand, just to make sure I still could. The smell of the lab, and, no doubt, my actions while I was there, digging through parts of the body that nobody was ever meant to see, seeps into my every pore, and my post-lab showers leave me scrubbing my skin, Lady Macbeth-like, trying to remove a smell that might at this point be entirely in my imagination. Out, damned formaldehyde.

The course has made me consider my own body in a new way. I have a newfound appreciation for my kidneys, my tiny, unassuming kidneys, lying in my abdominal cavity and dutifully keeping my entire self healthy and steady. I find the human body more impressive, more wondrous, now that I've seen its inner workings. It doesn't make me disbelieve in evolution, but in a way, allows me to understand those who do — how else could something so beautiful and complicated possibly have been devised by a series of random events?

I sometimes think about what it would be like if my body were the one being dissected, my identity reduced to a table number and an age at the time of death. The students would see my missing appendix, but they wouldn't know that the operation to remove it was the reason why I decided to become a doctor. They'd see my spine, abnormally curved and crooked — "kyphotic, lordotic, and scoliotic," a professor would tell them — but they wouldn't see the bulky, plastic-and-metal back brace I wore during my preteen years, as though I wasn't already awkward enough. They'd see my whole body, naked and defenseless, as a tool for learning about biology, but they wouldn't see all of the rest — the adventures it took me on, the late nights it saw me through or the torture it inflicted on me when I was a teenager and didn't think my body looked the way it should. But, they'd also see things I'd never know about myself — maybe an unusual branching pattern in my abdominal arteries or maybe a missing tendon in my hand.

Although anatomy ends next week, I won't quite be done with my cadaver — later we'll dissect her brain, which I'm sure will inspire a whole new set of existential crises and complementary nightmares. And, while I'll be glad not to feel the stench of fixative in my hair or worry about which foramina the branches of cranial nerves go through, I'm sad to be finished with the course, to accept that the body my donor gave has served its purpose, that it can finally be cremated and returned to her family. I hope they know how much she taught me, and how my future patients will benefit from the knowledge I gained from her. I hope that's what she thought about when she decided to donate her body, and maybe that it gave her comfort when she was nearing the painful end of her life. And, when her family gets her ashes, I hope that gleaming among the gray dust that used to be her flesh and bone, there are small, sparkling flecks of pink.

The Grand Tour of the Human Body

May 9, 2016

Krishna Constantino
University of Illinois at Chicago College of Medicine
Class of 2019

"**T**HIS IS A PLACE where the dead are pleased to help the living."
These words always greet me upon entering our anatomy lab. A similar saying in Latin is inscribed at the Palazzo del Bo at the University of Padua, home of the world's oldest anatomical theater. This site is witness to many of the lectures and demonstrations by Andreas Vesalius, the father of anatomy. This was during the time when human dissection was newly sanctioned by the church, but was only reserved for the corpses of criminals. Back then, human dissection was a privilege that was denied to medical students. *Hic locus est ubi mors gaudet succurrere vitae.*

The study of human anatomy had always been controversial. In the time of Hippocrates, it was Greek taboo for doctors to dissect a human cadaver. As a result, much of what Hippocrates learned was through animal dissections. It wasn't until Frederick II, emperor of the Holy Roman Empire, mandated at least one human dissection every five years that the science of anatomy was born.

Even today, I believe that the study of gross anatomy still brings about strong emotions. Before I entered medical school I had heard stories from the upperclassmen of people fainting or refusing to enter the lab. Well, if the dead were pleased to help the living, who am I to refuse them their pleasure? For me, there was something marvelous about studying the human body, peeling back each layer of fascia and muscle to uncover the vessels and nerves encased underneath. I have learned early on that despite all my hours of preparation for a dissection — reading the dissector over and over and carefully studying Netter's Atlas — no dissection turns out exactly as planned. There will always be an aberrant vessel or a nerve branch straying from its destined path and confounding medical students.

Despite the difficulties, I have to admit that I truly loved dissecting. For me, it was an experience similar to traveling. You study the maps, you read the travel guides, you look at the pictures of your destination and yet, nothing prepares you for the actual experience. The trepidation I feel each time I wield the scalpel, the fear that I accidentally cut an important structure, the thrill of discovering each studied organ just the way the atlas depicted, the confusion of a stray nerve or vessel supplying an unexpected area — these things are what I experience each time I go to dissect. It echoes Goethe's sentiments when he first visited Rome, the Eternal City: "Wherever I go I find something in this new world I am acquainted with; it is all as I imagined — and yet new ... I have had no entirely new thought, have found nothing entirely unfamiliar, but the old thoughts have become so precise, so alive, so coherent that they can pass for new."

I remember spending an entire afternoon dissecting the palm of the hand and, for once, the professor left my dissecting partner and me to our own devices rather than swooping down on us with the wrath of God each time we deviated from the dissector. His only admonition was to "make it look pretty." I had painstakingly scraped off the subcutaneous fat that was stubbornly clinging to my cadaver's superficial palmar arch — a graceful curve across the palm of the hand where the ulnar and radial arteries met — and was working hard at exposing the branches that radiated from the arch towards the fingers, the common palmar digital arteries. Each time I exposed a structure, I would feel a surge of excitement because it was exactly as the atlas depicted. That flash of recognition after spending hours looking at the pictures and imagining how it would appear took away my exhaustion. And so it is the same with traveling: you could look at photos all day and still find delight in recognizing something that you previously only saw in a picture. The destination is never a new place but rather a new way of seeing things.

Each traveler's vision of a tourist destination is never a single image but rather, an amalgam of multiple images. Our image of Rome is not a single picture of the iconic Colosseum but that of several pictures of the city. Similarly, in anatomy what the atlases illustrate is never the image of a single dissection but that of multiple dissections — each dissection forming the basis for the image of what the structure appears in the average person. However, anatomical variations are almost never depicted in atlases despite being present in a certain number of the population. I remember the first time I saw a thyroid ima artery, a variant artery arising from the brachiocephalic trunk. My professor had excitedly announced to the entire class to drop their scalpels and ushered us over to the cadaver with the aberrant vessel. Immediately, I knew that the brachiocephalic trunk didn't look right. It was quite a shock seeing four vessels sprouting from the brachiocephalic trunk instead of the three vessels that I had etched over and over into my brain. I could imagine that this would be an unpleasant surprise to surgeons cutting into the thyroid gland and discovering the unexpected artery.

I often have concepts of what things are supposed to look like and when confronted with the real thing, I am often forced to relearn what I thought I knew. Just as Goethe declared that "only in Rome can one prepare oneself for Rome," so it is with anatomy. No matter how much I studied those structures in books, it was only through dissection that I learned each structure intimately and how the pictures were transformed into real life for me. It is only by spending those hours in anatomy can I truly prepare myself for the reality of the human body in all its complexity, frailty, gruesomeness and magnificence.

A Donor's Story

April 10, 2017

Megan Kelly
The University of Arizona College of Medicine - Phoenix
Class of 2020

"In the end, people don't view their life as merely the average of all its moments — which, after all, is mostly nothing much plus some sleep. For human beings, life is meaningful because it is a story. A story has a sense of a whole, and its arc is determined by the significant moments, the ones where something happens. Measurements of people's minute-by-minute levels of pleasure and pain miss this fundamental aspect of human existence. A seemingly happy life maybe empty. A seemingly difficult life may be devoted to a great cause. We have purposes larger than ourselves."

—Atul Gawande, *Being Mortal: Medicine and What Matters in the End*

—

T O OUR DONOR,

The morning that we met was one most medical students eagerly anticipate as they embark on the journey that is medical education. Excitedly I put on my first set of scrubs, elated to look like a "real" doctor. Beneath my external façade however, I masked an underlying feeling of anxiety. *How would I react when I saw you for the first time? Would I be fascinated by the working machine that is the human body? Would I faint? Would I suddenly realize I'm not cut out for medicine?*

Our lab group gathered in a semicircle and stared intently at the cold metal table and electric blue zippered bag in which you rested. As our instructor made a few, now forgotten remarks and unzipped the bag, we peered over the edge and saw you lying there peacefully with your identification tag intact on your right ear. This unusual way of identifying a human being as just a number made the experience feel cold and impersonal. Our instructor quickly covered your face with a "privacy cloth." For weeks that was the last time we saw your face.

As we moved through each section of your body, dissecting away layers of fascia and removing major organs, I rarely stopped to think beyond the methodical cutting and scraping. While I had learned more in those few short months about the anatomical composition of the human body than many people have the opportunity to learn in a lifetime, I rarely stopped to think about you, the donor and person at Table 4. I did not know what your passions were or about the memories you had made throughout your living years; I did not know the meaningful story that was your life.

Just the other day I held your brain in between my two gloved hands. As I inspected the many sulci and gyri a realization came to mind: This was the organ that carried the facets of your personality, the struggles you faced, and the joy you felt. Prior to your final breath and allowing yourself to become a tool for education, you were a member of someone's family, perhaps someone's father or grandfather, a member of society. It was then that I truly began to appreciate the weekly six hours spent in lab, the countless nights studying and ultimately why every medical student has been required to take clinical anatomy since the infancy of medical education. Not only had I learned more about the human body than would ever have been possible from reading a page in a textbook, but I had also realized more about your story without ever having the privilege of conversing with you.

Over the course of four months we had inspected the various pathology that inevitably lead to your death. We learned that cancer had spread throughout your body, that your heart had to work harder than most, and that you had more surgeries during your life than our entire lab group combined. But we also learned that for whatever reason, you felt compelled to share your body and health struggles as an opportunity to allow students like us, discover how pathologic processes can lead to disease and how that impacts the complexities of the human body. One day, I know I will have a patient with cancer, who needs a kidney transplant, or requires a pacemaker and I will undoubtedly think of you as the first person in my capacity as a medical student, who experienced each of these health difficulties. As physicians, it our duty to look beyond the interesting and difficult problems our patients present us with and remember that every person we have the privilege of caring for is far more complex than the diagnosis they have been given or the ailment for which they are seeking help.

So to our donor, thank you. You gave me the opportunity to not only better my medical knowledge but also to gain a better sense of appreciation of the person beyond the pathology and acknowledge the moments that make each patient's story, theirs. I will be grateful to you throughout the course of my career for giving our table the opportunity to become more educated and competent future physicians and simultaneously more compassionate, understanding, and patient human beings.

With gratitude, Megan Kelly

Dead in Traffic:
Reflections on Gross Anatomy

February 21, 2016

Matthew Trifan
Lewis Katz School of Medicine
Class of 2017

> "I don't have a body, I am a body."
> —Christopher Hitchens, *Mortality*

—

*C*adaver. The word itself seems devoid of life. And, so too does the white plastic bag lying unceremoniously before me. It's the first day of anatomy, and I unzip the tarp and stare down at a wet, grey lump of clay. There it is. There is *what*, exactly? What was I expecting? Some warm human soul, freshly sprung from the loins of life? No. That's not this. The essence of life is gone — absolutely, irrevocably, unquestionably, gone. What's left is more like a husk of corn, or the molt of an insect. The head has been shorn to the thinnest veneer of grey stubble and crudely sewn back up, Frankenstein-like, with the brain removed. The dead-fish eyes are cracked half-open, like a glass doll's, with eyelashes curling daintily outwards and irises ghostly pale and alien beneath. The abdomen is bloated and moistened from two years of stewing in embalming fluid. Its skin is a sallow grey-yellow hue, with liver spots and freckles and moles faded to ashy dots, and small drops of fat beaded in the folds. The whole thing feels like leather left out in the sun — coarse, rubbery, thick, taut. When we lift it up to turn it over, there is an unsettling heaviness to it — "dead weight," I think grimly. And the nakedness? It doesn't register. In the postmortem lab, death makes sexless the human form. The breasts have long been flattened and become indistinguishable, and the genitals are squashed into an amorphous mass. Nothing about the cadaver's identity screams "man" or "woman." It's just *there* — a body, a corpse, a mummified relic for inspection.

In truth, I am not overwhelmed. My initial impression of the cadavers is one of mannequin lumplessness. Only over the course of the first week do they begin to assume a vague identity. I slowly uncover the telltale signs of human life in their cold remains — unearthing, like an archaeologist, the story within the ruin. The room begins to come to life. My eyes roam across the lab, and I spy on female cadavers lustrous red nail paint, as fresh and explosive as roses in bloom. The length of their nails — with their carefully manicured edges — unnerves me. Their strawberry sheen evokes images of old grandmothers making themselves up for one final visit from loved ones. "They want to look lovely before they go," I think, and something catches in my throat. Then, there are tattoos — a sailor with a ship on his arm, a woman with a rose on her hip and a man with a giant avocado on his thigh. Weird, bizarre images, all stamped with the spirit of human singularity. I find small, unspoken stories in birthmarks, scars, bruises, gold teeth and piercings. Depressions are found in the skin from watches and wedding bands. I see glimpses into lives I'll never know.

As the dissection continues, I unravel stories of adversity and the miracles of medicine. I discover pacemakers buried in chests — the testaments to failing hearts and second chances. I find holes in the abdomen for feeding tubes and staples down the sternum, where grave operations undoubtedly left family hands clasped in prayer. There are stents in clogged arteries, huge coils of springy wire found in the atrium of the heart, mastectomies, hernias, knee replacements, missing fingers, missing toes, screws in bones, corneal transplants, and cancer. God almighty, cancer: ugly, bubbling mountains of flesh spreading where there's no room to spread. Everywhere, I find surgeries and triumphs and tragedies chronicling the journey from *here* to *there* — from the walking, living, breathing, laughing lives of the deeply flawed, deeply mundane, deeply human to the silent, stillborn army in this lab.

Who are they, these dead in traffic? We know little about them: a single sheet of paper, no name, but an occupation ("pipe fitter"), an age ("82") and a cause of death ("multiple myeloma"). However, it's enough of a story for some of us to form a bond of our own, a doctor-patient relationship of the oddest sort. To an outsider, this might sound peculiar — a relationship with a dead stranger. But this is no stranger. After weeks and hours of working with one body, meticulously cleaning its muscles and organs, that cadaver becomes a quiet companion, a familiar face. Our friendship is born of the knowledge that these donors willingly offered their bodies for the purpose of being examined, exposed, cut, chopped, skinned and prodded. They must have understood how little dignity would be afforded to them. They must have known that our scalpels would be blind to pride and vanity, and that, when we finally finished, their body would be disfigured beyond any recognition of their former self. All that would remain, would be ashes, carefully collected, cremated and commissioned to their loved ones.

And yet, I feel, this is what *they* wanted — to disappear entirely, like a candle burning brightly from both ends. And, in doing so, they have illuminated my world.

I try to cling to that sentiment as I face the dirty, grisly reality of what I'm doing. What other choice do I have? The noble field of medicine is built upon the grotesque and the unsavory. We embark on this profession with the grisliest journey: the dismantling of the human body. In the crudest terms, this means picking apart every bit of sinew: skinning a face, sawing off a leg, transecting a labium, prodding and poking and opening a heart, a brain, a lung, a bowel. Yes, it is hard to think about. But, as the days and weeks roll by, these tasks become easier and quicker, bred with familiarity of the body and an engrossing fascination in the work itself. I become keenly aware that most people will never have the opportunity to do what I am doing, unveiling the mechanisms of life, intertwined beneath our skin like the inner workings of a clock — why should it not be enthralling?

Of course, when I talk to my friends about dissecting a human body, they recoil in horror. Their minds go to dark and ghastly places, spurred on, no doubt, by sadist television and silver-screen cannibals. Not only do they bear a physical abhorrence to the act, but an emotional repugnance at the desecration of flesh. And why not? The sanctity of the body and spirit has spanned the arc of history. For hundreds of years, the Church maintained that human dissection was a desecration of the human soul. Until the 1800s, punishment for violent criminals often included "medical dissection after death," an added horror to those believing their immortal souls were at stake. Even today, the human body is seen as sacrosanct. For many of us, our sense of "self" is intricately bound to our physical essence. Perhaps we believe that we may, in one form or another, live on in the physical world long after death has claimed us. Perhaps we are searching for another form of immortality. "From my rotting body, flowers shall grow," wrote the Norwegian painter Edvard Munch, "and I in them, and that is eternity."

But there are other ways to live on in the world: namely, in the art of knowledge, or in the passing of ideas from one generation to the next. *Mortui prosumus vitae*, reads the inscription on many donors' tombstones. *Even in death do we serve life.* I find it such a beautiful sentiment to serve as a teacher from beyond the veil of death. I liken the feeling to that old children's book, *The Giving Tree*, in which the tree has offered everything to the boy it loves, but discovers one last gift from the shadow of the grave: an old stump for sitting and resting. There's something wonderfully cyclical about the lessons of life and death to be found here: something that catches at the essence of religion, philosophy and art, some final message passed on to us by our elders, who formed us within themselves. From the dying, we learn how to live. And likewise, "While I thought I was learning how to live," wrote Da Vinci, "I have been learning how to die."

It's hard to feel sorry for my cadaver, when I'm just a few quick dreams from his place. He is the future, the unyielding truth. When I look at that cold metal slab, I see myself there — empty, undreamt, and unborn. Shorn of pride. Free of self. Monastic, destitute, naked and fearless. Surrendered against life. And that's okay. That's comforting, and peaceful and beautiful.

13

It's the completion of the circle we were born into. My only question for the dead man lying before me is this: Did he learn to let it go? And will I?

For me, that journey begins with a cadaver, a scalpel and a nervous heart.

The First Cut

November 3, 2016

Nyembezi L. Dhliwayo
Rosalind Franklin University of Medicine & Science
Class of 2018

I took the first cut,
And suddenly it hit me ... A wave of emotions
At once exhilarating, at once terrifying
A rush of adrenaline, a rush of shame
A rush of excitement, a rush of panic
A loud gurgling from the deepest pit of my
gut
I had to look away

I took the first cut
My mind lurched and screamed,
so silent, yet so loud
Here lies someone's mother, someone's
grandmother
So real, yet so lifeless
Everything about her so alive
Her arms, her wrinkles, her freckles, her
breasts,
The wisps of hair trailing off into the
deepest of her private parts
So exposed

I took the first cut,
Resisting the urge to reach out and shake her
To tell her she's the first dead person I've
ever seen!
Wake her out of her deep slumber,

Get her off the cold metal table
And ask her what it took
To live 94 long years

I took the first cut
She has beautiful arms they said,
Awesome muscular definition
Not too much fascia
She must have lived a healthy life no doubt
Yet here she lay at the end of it all

I took the first cut
Everything stood still!

Overwhelmed, anxious, excited
I'm completely connected,
Conversing intensely with her;
Fate has brought us together I tell her
Your body my willing teacher
And me your student,
Eager to learn
To explore the human body
So humbling

I took the first cut
I am on course to becoming a healer
But first I must encounter death
So ironic...

I took the first cut,
Nothing could have fully prepared me for this
I take a deep breath
Plant my feet and continue dissecting.

Careless Curiosity:
Reflections from the Anatomy Lab

May 7, 2015

Jennifer Tsai
Warren Alpert Medical School of Brown University
Class of 2019

O N MY LEFT, THE ragged, meaty stump of a severed neck stands upright like an abandoned signpost. A classmate examines it carefully. She is petite. She stands on a stool to obtain a better angle for observation. She cranes her neck, twisting the very muscles she is studying back and forth, back and forth.

To my right, a peer brandishes a hacksaw, ready to bisect a man's face so it falls open like a cleanly sliced hardboiled egg. At the table next to me, a friend steadies a bone saw upon his cadaver's skull. He kisses blade to occipital bone — gentle — furrowing his brow as bone dust leaps like a celebration.

I'm snapped out of my reverie by the approach of our lab guide. She bounds cheerily, excited by our progress. "Looks great!" and then, "any problems?" I look at our cadaver. His head lolls forward, spilling into a wide open chest cavity, nose resting delicately on his own exposed liver, spinal cord flayed open, skull cracked and sawed apart. The skin of his face has been peeled away by our scalpels, brain removed and sliced. His neck is a tangle of flayed muscle fiber, viscera separated from backbone, shrunken eyes glaring sideways, both in shame and accusation.

He is desecrated. Flung apart by curious fingers and obedient steel blades.

"Just having some issues finding the demarcation of the pharyngeal muscles," is the answer my lab partner gives.

She raises her voice to be to be heard above the constant thwack-thwack-thwack of mallet and chisel, the whirring of bone cutters, the deadened zip of hacksaws flying through skull. Added to the cacophony is the occasional rip of viscera and connective tissue as it is torn off the muscles it has sweetly embraced for decades. The noise swells with the excited chatter of my peers. Formaldehyde grips our nostrils, coats our skin like a thick paint.

I am reminded of how the white coat can normalize an exceptional amount of violence.

It is heavy in the air yet it does not always weigh on our minds.

Perhaps the most poignant moment of anatomy lab is the moment we expose the cranial nerves. They look as if they were painted with the thinnest of brushes, tattooed like fine meshwork in clean white ink. Beautifully spun. Delicate. Tiny tendrils of white matter that spill across the brainstem.

Cutting the cranial nerves, more so than any other act exerted upon this body, feels most acutely like I am destroying the legacy of this man. The nuances of his being are falling away beneath indifferent, deadened steel. We learn that these fibers carry instructions to shrug shoulders, lift soft palate, curl tongue, smile, laugh. Such thin threads, impregnated and fat with the implications of life.

And then the destruction. One lick of the scalpel edge, and there. Severed.

Curiosity is a beautiful thing, but sometimes in anatomy lab, it feels ugly.

The only way I can accept this man's sacrifice is by destroying the gift he proffers. We dig for discovery and then discard. Hold, but do not keep. We appreciate the anatomical structures, but do not give thanks. We accept his donation by pillaging it. It is an intimate exchange despite, or perhaps because of, the brutality of our damage.

I cannot help but to feel disrespectful every time I enter the anatomy lab. We proceed with every session as if we are entitled to these bodies, just by virtue of being medical students. I think about how little we are reminded of our privilege. I think about how, without our white coats, this activity is a felony. I think about how the person on this table loved and was loved, and how we flay his body in the name of academic exercise. The sight of medical students huddled around carved open bodies continues to conjure an image of consumption. Of an intellectual cannibalism.

Our curiosity often feels careless. We do not practice surgical techniques or indeed engage with any part of healing. We are instructed to rip, tear, probe, and slice muscle and mesentery, with no threat of consequence. We invade with little caution or fear.

We cut, look, learn in the hopes that this knowledge will help others, but at this stage of our medical journey, our discovery is purely for our own benefit.

Anatomy lab is a strange source of personal profit.

Anatomy as Art: Installation #12

May 30, 2016

Nita Chen
Albany Medical College
Class of 2017

For the majority of medical students, gross anatomy is the first time that we observe and cut into the flesh of preserved cadavers. Whether it is through a longitudinal year-round program or a semester's worth of concentrated anatomy, most of us develop a unique relationship with the cadaver gifted to us by generous donors.

At Albany Medical College, upon our orientation to gross anatomy, we are asked to draw our feelings on blank index cards prior to entering the cadaver laboratory. As we progress through the year, our sentiments regarding anatomy may remain the same, or may change, and these drawings allow us to look back at this milestone we crossed as budding medical students.

There is no substitute for experience. No matter how many times you read about how to do something or how many simulations you participate in, nothing quite captures the essence of a true experience other than the experience proper. This concept relates well to the first encounter medical students have with actual patients. Though they are usually well-versed from a conceptual standpoint by the time they see their first patients, most medical students feel that they learn and relearn so much when they collect their first real histories or perform their first real physical exams. For many of us, our first patients leave a lasting impression, shaping the way we provide care. They also demonstrate how important relationships are in health care, especially in a day and age where technology dominates and people often feel disconnected from one another.

First Patients

Pupil of the Eye by Kaitlyn Dykes
Class of 2019 at Sidney Kimmel Medical College

Seeing Medicine

January 27, 2016

Shivani Srivastav
New York University School of Medicine
Class of 2018

I LEARNED TO SEE THE world through words. Words I picked over all day at school then curled up with long after I was supposed to be asleep. I reflect through writing, turn to all 1124 pages of my worn *The Lord of the Rings* in every difficult time, and dream best with my head pillowed on a book.

Yet I never realized how fully these words and stories shape my experience of the world until I entered medical school, a space where the world is not often viewed through the lens of stories. In college, I took pre-med classes as a means to end: becoming a doctor. However, as a creative writing concentration English major, I really flourished in my writing courses. I immersed myself in the minds of book characters and authors, learning to understand myself and the people around me through fictional beings. My English courses were the land of "no right answers," where every discussion centers on a "why" question. Why would a character take a certain action? Why would an author describe the character this way? Stories and poetry create worlds of ambiguities and shifting perspectives, inspiring and challenging each other. To appreciate these nuanced worlds, one must pay attention to details: their building blocks of words and punctuation.

Like English, medical education requires attention to minute details. Take the stark white on black of X-rays or the gradients of darkness in ultrasound. First, we learn which parts of the body will show up as bright or opaque, what it would mean for a line to be sharp or blurry. Then using this information, we build an image of the body. Suddenly, a shadow in the ultrasound field, one I may not even have seen a second ago, is called a gallbladder. There is often strange dissociation between how these body parts are viewed in medical school versus the "outside" world. On the outside, blood gushes forth red and panic-inducing. But in my life right now, blood exists as the shades of pink and blue polka dots on pathology images. Those tiny cells, smeared onto glass and fluorochrome-stained, now huge on the projection screen. This language consists of shape and color: which cell is pinker, denser or has nuclei off-center? We use these details to name what we see, normal process or disease.

This progression from learning to read these images to understanding them seems linear, yet, for me, it's fraught with difficulty. While reading stories, I easily build mental bridges from the mechanics of language to its grander purposes. I savor words and turns of phrase for pleasure. I expand my view of the world by being in someone else's head. I travel, crossing continents or scoping out imaginary societies. I dissect stories, interpreting those building blocks of style, word choice, even spelling — but all this leads towards understanding, towards absorbing the words and applying them to the world I live in. Whether my sense of justice, honor, spirituality or love, each has been filtered through the words of countless authors. I miss this in medicine. I learn to read CT scans and X-rays and turn their myriad shades into coherent diagnoses. Yet as I do so, it remains hard for me to build to their grander purposes.

After all, learning to navigate these multivariate languages of medicine is a feat in itself, one our medical education revolves around. One day, we may be studying the (literal) Latin of anatomy terms. Another, we may start listening for the sound of an irregular heartbeat. With all the effort learning each of these skills requires, interpreting them becomes an end in itself. This cloudy gray shadow is a pleural effusion on X-ray. That bright spot is a tumor on MRI. But why are we learning these things? If I am looking for a larger meaning, then I am seeking the reason behind these diagnoses. Where is the patient in our medicine?

Stories draw me closer to people: author, narrator, character or — in happy times — someone else reading the same work. Yet, in medicine, we learn many of our skills in a vacuum. Eventually we may read an X-ray to help treat a patient, but right now it does not apply to anyone needing treatment. The pathology slide we examine has no connection with the full constellation of a person's existence, and perhaps it never will. After all, we are not expected to interpret every kind of test once we graduate medical school. We are destined to subdivide and specialize. Pathologists will immerse themselves in their slides, radiologists in their scans, and so forth. So, what does learning all these methods teach us?

I hope to learn from these medical methods again what reading taught me without my knowing it: there exist countless ways to understand people. A poetic stanza, a phrase of prose, a word in a foreign language or one made up — each exposes humanity differently. As doctors, we are meant to treat our patients, to heal their bodies. To do this and understand someone else's body, we examine it with our tools. X-rays, ultrasounds, MRIs, physical exams, blood tests or pathology slides — none is an end in itself. Each is a different lens placed over the body to show us something not otherwise apparent. I hope we recognize these lenses as such: mechanisms refracting a portion of a patient's reality for us to see. None can reflect a patient or her body in totality. Instead, each becomes another way we could learn about ourselves or those who surround us, another story we can read.

Wearing the White Coat: My First Preceptor Experience

October 20, 2013

Qing Meng Zhang
Rush Medical College
Class of 2017

M Y HANDS WERE A bit sweaty; my heart was fluttering. As I was driv- ing, questions and doubts surfaced in my mind. "What if my preceptor is mean?" "What if I put my stethoscope the wrong way?" I had shadowed physicians and worked with patients in hospitals, but this was different.

This was the first day when I would become a part of the medical health professional team and would utilize the physical diagnosis and patient in- terview skills I had learned in class. This was the moment where I would wear my brand new white coat decorated with all sorts of gadgets: my med- ical student ID, stethoscope, reflex hammer, pen light and notebook.

I walked into the clinic and talked to the receptionist, who instructed me to wait. Feeling slightly awkward in my crisp white coat, I sat in the crowded waiting area. Instantly, I was greeted with gazes of a mixture of curiosity, intimidation and (maybe) respect, as well as some friendly smiles. Wanting to blend in with the crowd, I secretly wished that I had waited to put on the white coat, which felt strangely foreign to me. Underneath, I was just the same as everyone else in that room, an ordinary person with very little medical knowledge. And yet, the mere piece of fabric has put me in an entirely different position. I felt even more nervous.

I was eventually called in by my preceptor. After seeing two patients with her, I asked if I could take the next patient's blood pressure. Although I was confident in my skills, I tried to maintain my self-assurance as I at- tempted at the stubborn sphygmomanometer that would not inflate.

Is it hooked up properly? Yes. Is the knob closed? Maybe. Righty tighty, lefty loosey. Okay, try again. Nope. Maybe use a different one. This is really embarrassing. This is taking a long time. Oh, okay, it's finally working.

Satisfied with my reading, I proceeded to record the rest of the vital signs: bilateral radial pulses, respiration and temperature. When the physician left the room briefly, I began conversing with the patient about the idiopathic pseudotumor in her brain, as well as her family and career plans. She expressed her frustration that her doctors could not provide a definitive answer to the cause, nor could they agree upon a treatment other than to manage her symptoms. The darker side of medicine she exposed mirrored my observations throughout my clinical experience prior to entering medical school. Nonetheless, the patient was optimistic and showed a great appreciation toward my preceptor at her understanding, honesty and open communication.

A knock on the door. The next patient was ready, with new stories to hear, new people to help, and new diagnoses to learn.

As medical students, we are constantly bombarded with seemingly endless amount of medical and scientific knowledge to digest. We work diligently for the day when we can see and treat patients independently and make a difference. In the process, we learn how to think like doctors and speak in medical jargon. But while being buried by piles of notes and drowned by stress, engaging in various clinical activities reminds us of the purpose of our struggles and hard work during medical school. It is a reminder that we, the students and future physicians, are just as human as everyone else, and need motivation and hard work to succeed.

Ensconced in His Throne Upon the Sea

January 31, 2018

Ashten Duncan
OU-TU School of Community Medicine
Class of 2021

Sonnet I: A Traveler of the Sea

I strode down hallways, winding 'round to meet
A sailor old and take to him his meal.
A gentle bounce in every step on beat,
This home to many always builds my zeal.

His room I entered — knocking first of course —
And said then, "Mister Harold, pleased to meet!"
He turned to greet me, smiling, voice so hoarse.
"Hello, hello, my child! What's good to eat?"

I spoke with him as fast an hour elapsed.
I heard his "hist'ry," all about the sea,
Before he sat to finish his repast.
I queried Nurse Noëlle of things to be.

"A textbook glioma consuming his brain,"
She sighed. "One that now will not rest, nor refrain."

Sonnet II: Ashore with His Child and Wife

He raised his head so weak and smiled at me
As I approached the man I came to serve.
"It's great to see you, young'un," perked up he.
"A lovely day," observed he, full of verve.

His lucid state — relief for those near him —
And speech intact did make me feel at ease;
This sailor old was dear, a priceless gem.
"You know, I miss the warmth, the light, the breeze."

He asked about my dreams and future road.
He asked about my quest for doctor's oath.
I answered blithely, spoke of seeds I sowed.
"My family guided me, allowed my growth."

"I see your shore: a bright and wondrous life,"
He beamed. "Like my own child and loving wife."

Sonnet III: Waning Vitality, Waxing Courage

Delir'um boasts a state that's hard to stand:
Its veil concealing all the face and soul.
I greeted him and shook his clammy hand.
He glanced at me — the cancer taking toll.

I took my place with him at his bedside.
I saw it: clear disease process at work.
His vacant stare did chill me, deep inside.
A pang of sadness stabbed me as a dirk.

He rubbed his head with such a trem'lous hand.
The tumor grown and drugs galore did cloud
A mind once sharp, alive, profoundly grand.
They stripped his vigor, drained all his endowed.

"Don't worry, Mister Harold, I'll return.
Tomorrow, six o'clock, we'll chat and churn."

Sonnet IV: Asylum for the Caring

Away was I when fast his breathing ceased,
Away at home when ere he did contend.
A burning candle marked a life released
More than a cerement ever could portend.

The vigil over, knew I he had passed.
A nurse so kind embraced my shoulders cold,
And there I stood, still caught within a blast.
She handed me a candle lit to hold.

We solemn few then entered hallowed space
And soon prepared his form to be retrieved,
Concealing well his vessel left, his face,
Beneath a sheet opaque — the sleight achieved.

That evening moon, that candle's flame,
The peace I felt, an empty room now tame.

The Value of Empathy in Medicine

March 6, 2015

Sarah Bommarito
Wayne State University School of Medicine
Class of 2016

EMPATHY: IT'S WHAT SUPPOSEDLY drives us to become physicians, and what we're told to demonstrate through our extracurricular activities and during our interviews. We yearn for that perfect patient interaction in which we comforted or understood in a way that changed the patient's perspective on medical care. In our idealized view of medicine, we truly believe that empathy will be our saving grace throughout medical school, residency and beyond. If we can simply connect with our patients, then we will succeed and the patients we care for will thrive.

And then we begin our clinical rotations.

During orientation at my clinical site, one of the attending physicians gave a presentation on how to avoid becoming "robots" during the course of our medical training. I sat in the audience and wondered how I could ever fully disregard emotion, but the entire premise of the speaker's presentation hinged on the certainty that this would happen unless we tried very, very hard to remain human. I considered this for a while and decided that while I could understand how some of my classmates might lose empathy, my hyper-empathic tendencies simply could never be diminished. After all, studies have shown that physician burnout is worsened by forcing down negative feelings that naturally occur over years of bending to the burden of patient care.

In the following days and weeks I kept this lecture in the back of my mind as I began to learn how to interact with difficult patients, take overly detailed histories, complete monotonously thorough physical exams, and present this information to my often-impatient superiors. Early on I realized that when you're part of a medical team, empathy isn't valued nearly as highly as virtues such as efficiency, confidence and medical knowledge — and did I mention efficiency?

Many aspects of clinical care can be reduced to measures of time. Insurance companies become vocal if they have to pay for a longer length of stay. Patients become irritable if you ask seemingly similar questions too many times — even though we've all seen that slight rewordings of a question can yield strikingly different answers during rounds. The interns to whom you're assigned are too swamped with paperwork to care about the fact that your patient is apprehensive about being discharged and wants more information before leaving. The attending physician has to be somewhere this afternoon, and how could you even consider bringing up patients' trivial concerns when there are lab values to be discussed and additional tests to be ordered before lunch?

It isn't that one part of the system demands efficiency — it's that most parts do. In order to function as a valuable member of a medical team, you almost have to conform to these standards. People and companies expect this from you. You learn to optimize the amount of work you can do and the number of patients you can see in a morning and an afternoon, and the standards continue to become more rigorous.

During my second rotation, I was placed on an oncology floor. In the beginning, I often inconspicuously shared in the sense of sadness and hopelessness that my patients and their families faced when trying to understand their diagnoses and prognoses. I learned not to go to my superiors with details of personal talks with patients after being told and reminded that I wasn't allowed to be sad after long discussions about impending death. It was my job to be strong for patients so they wouldn't see tears in my eyes and assume that all hope was lost.

Within a few short weeks, I had accepted that stage IV cancer could be equated to certain death in a shorter amount of time than any patient was ready to accept, and I had even begun to think somewhat negatively of patients and family members who stubbornly refused to agree with these now-obvious realities. In large part, this way of thinking shielded me from the pain of relating to these patients. I began to think that medical professionals might be on to something: maybe it's right to disregard empathy in favor of a more logical approach to patients, a bedside manner that is more realistic than overtly comforting.

However, I was forced to reconsider this idea. This way of thinking caused me more exhaustion and misery than I had ever felt in my medical training. I went home with headaches, complained about my day and then holed up to study. While I enjoyed learning about medical conditions, I felt little emotional pull to continue my day-to-day work.

And then a patient and his family put me back on track. The patient, middle-aged and in a relatively new marriage, had been diagnosed with stage IV cancer about two weeks before I first met him. When we first spoke, he was irritable and quick to disregard me. His wife apologized for him several times during the difficult interview. I told her I understood that he was in a great deal of pain and exhaustion and appreciated their patience with me.

Over the next few days, she began to tell me details of her husband's illness that never came up during rounds.

The patient declined slowly at first and then, after a routine procedure, his mental status became severely altered. No available treatment could prevent further decline. The team recommended hospice care to keep him comfortable. His wife, though upset, seemed to understand, but admitted that she needed time to think and to discuss this with their immediate family.

By the time I went to the patient's room to check on him later in the day, his wife's attitude had changed drastically. She began to accuse the team of giving up on her husband. She told me that he was a good man who had provided for his kids for many years. He was a hard worker. He cared so much about her. Why did he get worse so suddenly? She wasn't ready to give up on him; why was our team giving up?

I listened to her with conflicting emotions. I knew this was difficult for her, but I also knew from my admittedly small amount of experience that the best thing to do was to ensure his comfort in his final days. But something in this woman's tone sank into me and reminded me that we were dealing with a unique patient. She hadn't faced this situation before. This is easy to forget after seeing so many patients with similar afflictions. Even in the small amount of time I had been on the service, I began to comprehend that physicians become used to certain situations despite changing patients. Disregarding a patient's individuality allows for treating the condition based on medically sound evidence without the need to understand the patient's whole story. This certainly saves time and energy that may be better spent on other tasks, after all.

At the end of the woman's tangential musings she began to cry — quietly at first, then with sobs that shook her body. Without considering my actions, I moved a box of tissues closer to her and placed my hand on her shoulder. I told her that we weren't giving up on her husband, and that we would do everything in our power to decrease his pain and maintain his comfort. I told her I knew her husband was a good man just by seeing how much she and their family cared for him, and that the team would be there to answer her questions and do as much as possible for her and her husband. My sentiments were genuine. Nothing I could have said would have soothed her entirely, but she was noticeably less upset and more open to considering the difficult decision at hand.

Empathy didn't save the patient's life, of course. He passed away comfortably under hospice care a couple of weeks later. But that day — in that moment — I knew I had improved the situation simply by seeing the patient's humanity and relating to his wife. While our conversation did not make the decision for her, it seemed to give her a sense of peace in making a difficult decision. Because of her decision, her husband was ultimately granted comfort in his final moments.

A later experience during my psychiatry clerkship provided further support for the value of empathy. I was placed at an addiction treatment facility for two weeks and primarily saw outpatient encounters. One day, I was invited to a group therapy session in which the patients wrote letters to their addictions, and the residents and students wrote letters to those suffering from addiction. I wrote about how I hoped patients would find support, care, comfort and acceptance in us and how proud we are that they were taking the necessary steps to find clarity and overcome their addictions to lead healthier, more fulfilling lives. In short, I told them we're on their side.

While I wasn't able to stay to hear many of the letters, the therapist later told me that when she read our letters to the group, the patients became very emotional. They hadn't realized how much we cared, she said, and it changed their perspectives on seeking care from physicians after their stay at the facility. Many of them had felt discouraged at their appointments and assumed that we saw them as a nuisance. In reality, many of us saw them as strong and determined people who were ready to get back on track and repair their damaged relationships and lives through treatment.

Showing empathy to patients can mean more than we realize or understand. Simply knowing that someone truly cares can help patients feel comfortable and justified in seeking the appropriate treatment or making a difficult medical decision.

As medical students, we're in the unique situation of experiencing some amount of responsibility to care for patients as well as the time to do so in a way that many residents and attending physicians cannot. Our time isn't billed to insurance companies, and there are so many options when it comes to how we can choose to spend it. I realize that I will reach a point where my time will become more limited, when I might find myself in a more robotic state with the eightieth patient who presents with a certain condition, but I hope I find the strength to return to this empathic state of mind at least as often as I leave it. I hope no one is ever able to fully convince me that empathy in medicine isn't worth the effort.

In continued lectures, I am reminded that empathic interactions need not take more than a couple minutes to cause a positive impact, and I've seen that a couple minutes is often enough to show a patient that you truly care about his or her outcome and that you are undoubtedly on his or her side. Sometimes patients and their families need to feel that sense of shared concern in order to feel comforted and adequately cared for. Sometimes physicians need to remember that their patients are human in order to avoid a sense of emptiness in their careers.

I am often given cause to wonder, even if the interaction takes slightly longer, what do we have to lose by showing concern, listening to our patients, and allowing ourselves to hold on to that feeling that drove us to become physicians in the first place? In truth, it seems that, to the contrary, we all have so much to gain.

Do You Remember?

March 5, 2014

James Yan
Schulich School of Medicine & Dentistry
Class of 2015

THERE EXIST, IN TRUTH, three simple words that strike dread into the hearts of every physician:

Do. You. Remember.

This phrase was introduced to me in the middle of first year. I was spending time in my medical student lounge when a link popped up on my newsfeed to a TED talk by Dr. Brian Goldman, an emergency physician from Toronto who hosts the radio show White Coat Black Art and who has also authored the book *The Night Shift*. In the video, Dr. Goldman reaches out to the audience about changing the medical culture of silence around making mistakes. (Check it out if you haven't seen it yet, it's worth the break.)

When I watched the video of Dr. Goldman speak, it was clear to me that he was haunted by the encounters that left him remembering specific patients.

Fast forward two years. I'm in a family medicine clinic, trying to learn the ropes of ambulatory medicine. A lot of the exposure here is the "bread and butter" cases: sore throats, fevers, rashes, well baby checks, immunizations — all good for honing in on a consistent approach. The work done by clerks involves history and physicals, contributing to an assessment and management plan, as well as conducting basic procedures. To assure patient safety, we're supervised and things are double- and triple-checked by residents, nurses and the attending physicians. I assumed it was a pretty safe system.

Yet one Wednesday morning when I was just arriving to the clinic, Angie, the team nurse pulled me aside. She had a phone in her hand.

"Hey Jim, *do you remember* Greg Foster?" Her voice was hushed, almost solemn. "He came in yesterday with the sore throat."

Yes, Mr. Foster, 45 years old. He had a two-week long sore throat and a cough that worsened over the course. He stated that the cough did not bring up much sputum, but did produce a sparse, thin and white mucus. No fever, no shortness of breath, no chest pain. No nocturnal symptoms, but he did feel his voice was hoarse. He had not travelled recently and never smoked. On examination he appeared well, not dyspneic, with an occasional dry sounding cough. His heart sounds were present and normal, and his lung fields sounded clear. Pretty much unremarkable up to his pharynx which did look red and irritated but without any swollen glands or petechiae. All vital signs stable. As per the protocol at our clinic I brought in the attending who conducted his own exam. Satisfied, he suggested I perform a rapid strep test but predicted, correctly, that it would be negative.

"It's good for you to practice doing these things," he said, as Mr. Foster was amenable to the idea with a trainee practicing on him.

In the end we thought he had a case of viral pharyngitis and he would soon recover with some more rest and symptomatic management. Mr. Foster seemed reassured and left, and I logged the case, which the resident also reviewed and signed off on. Later my attending reviewed some clinical pearls for sore throats: the group A *Streptococcus* pharyngitis guidelines, red flags for more serious conditions, and which diagnoses you want to make sure are ruled out (such as peritonsillar abscesses, cancer and epiglottitis). This is pretty typically how the cases went for me: straightforward.

So why was Angie asking if I remembered Mr. Foster? A low rumble of dread swelled in the back of my head. *What did I miss?*

"Yes, I remember him. He was here yesterday with a sore throat."

"Yeah, it's his wife calling, she's very upset because he was rushed to the emergency room last night, unable to breathe. He had to be intubated at the hospital and he's stabilized now, but it was dicey for a time. The docs at the ER said that it's epiglottitis, and that we should have caught it earlier."

My heart sank. My mind raced. Thoughts exploding out at once. Many of them expletives.

Frantically, I tried to go over the whole encounter in my head again and analyze it again from another angle. Maybe I could recall what was missed.

Was it the hoarse voice? It had to be an early sign, how could I have overlooked that! Stupid stupid STUPID.

My attending came and settled the issue by stating that this "miss" was not something we could have predicted yesterday. Furthermore, I need not think of this as my fault, as there was a whole team of people that were also involved in Mr. Foster's care. Yet for the rest of the day, I was shaken.

I think I understand why physicians hate that phrase. It triggers the mind to become obsessively masochistic. It opens Pandora's Box: doubts in one's abilities, anxiety for the subsequent consequences, and fears for the patients. Once opened they can run rampant. The demands of expecting perfection and being immaculate adds to the stress of facing errors. Phrases like "we should have caught it earlier" implicitly throw the blame on other physicians and perpetuates the ideation that mistakes happen only for "bad" doctors.

At the same time, it's irresponsible to shrug off mistakes nonchalantly by assuming that these incidents are inevitable. By doing so, especially at the trainee level, we not only lose out at a valuable moment to learn but more so, fail to appreciate the outcomes of our actions on the patients. Working as a clinical clerk is not a sterile bubble for us to just learn and "play doctor," but puts us in an environment to start to learn how the consequences of medical decisions can unfold.

It takes experience to find balance within these moments, when we are reminded of our fallibility, to remain assured in our abilities, yet still humbled by the incident to learn. As a student, I have to accept that mistakes and errors will happen and I should not become paralyzed by my fear of making them.

Do you remember?

Before, I wondered when I would first hear this question, and questioned how I'd react to hearing it.

Now, I guess I'm wondering how many more times I'll hear it before I graduate.

Personal challenges are certainly not unique to medical students, but the influence medical school has on these experiences is profoundly different. Consider how, for example, an illness in a family member evokes thoughts of patients encountered during clinical experience in Aleksandra Bacewicz's "Grandmother." This cognitive phenomenon is common for medical students. Other challenges represented in these selected works relate to cultural identity, mental health struggles, and the medical student's peer group.

The pace of medical school can prevent many students from paying serious attention their personal life as they often prioritize their academic performance above all else. We have seen that writing allows many students to find comfort in, and even resolve, some of these personal difficulties. If that's not the right strategy for the medical student in your life, we encourage you to start a discussion to identify alternative approaches.

Personal Challenges

Wistful Thinking by Sachi Gianchandani
Class of 2018 at Oakland University William Beaumont School of Medicine

Personal Challenges

One Step 1 Experience

August 16, 2015

Nathan Juergens
University of Minnesota Medical School
Class of 2017

TICK-TOCK. TICK-TOCK.
The only clocks in the room were the digital, silent type, but still I heard it.

The first hurdle to becoming a board-certified physician was looming as the ticking in my head grew louder.

It is now the summer following my second year of medical school at the University of Minnesota, and students across the country have just taken the eight-hour-long, 308-question United States Medical Licensing Examination Step 1. Here I outline some of my experiences preparing for and taking Step 1.

The Importance of Step 1

Residency programs use Step 1 scores when deciding who to invite to interview at their program and hospitals. An individual's score is as important as any other factor involved in this decision, be it research, extracurriculars or clinical grades. The weight of this didn't fully settle onto my already bowed shoulders until well into the six-week-long study period. How hard I worked then, how well I did on this multiple-choice assessment, may change what I end up doing on a day-to-day basis for the rest of my life. Even if the difference between an average and a good score doesn't change the specialty I choose, it could certainly change the location of my residency, the city I will live in for three-plus years following medical school. Silly mistakes and poorly-studied material would have more influence on the course of my life than ever before, as would the lucky guesses and extra nights spent committing information to memory. Taking time away from studying became more stressful than the studying itself.

The Good Ol' Days

A calendar year ago was the last built-in summer break of our academic lives. This was also when we started hearing murmurs about the Step 1 exam. Extra-institutional board resources started to pop-up in conversation. Many students bought Pathoma, a succinct review of all of the pathology we would study in our second year, written by Dr. Husain Sattar from the University of Chicago. Along with First Aid and the USMLE World Question bank, Pathoma is part of the Holy Trinity of Step 1 studying, and a good book to have before getting into the second year organ systems courses, which are the real meat of medical school, so to speak.

That year started with the proverbial pedal pressed down to the underlying metal. Time for kidding around was quickly evaporating. Heart and lungs first. Blood and guts second. Brains and nerves third. Winter break. Exhale.

I liken winter break of second year to warming-up before an intense sporting event. Nothing had really started yet in terms of Step 1 study. But some students were doing a few wind sprints. Others were casually stretching out their hamstrings and talking to a coach. Some were getting a long drink of water and watching others warm-up. By winter break, Step 1 was on the mind.

And then we were back, covering renal, endocrine and reproductive health. By this point many were using Pathoma regularly, as well as First Aid, the Bible of Step 1 studying (or whatever holy book you prefer). Everything we need to know is at least mentioned in this text, but maybe not to the depth we need to know it. It is a colorful book, dense with information and helpful mnemonics. It's really a joy to read, if it didn't have such an ominous association, because by the time you are going through and understanding First Aid, the End of Days is nigh.

The Countdown

Outside our windows, winter gave way to spring, but inside it felt more like fall transitioning to the dark days of the coldest season. The final unit, Human Health and Disease 5, was upon us. In each of the previous four units, organ systems were paired that interact with each other, or have overlapping learning points. Blood flows from the heart to the lungs to receive oxygen, before being pumped into the rest of the system. The brain is obviously the focal point of psychiatric illness, and is made of the same general tissue as peripheral sensory nerves. Diabetes is an endocrine illness that often manifests as kidney disease. This fifth and final unit, however, was more of a grab bag of information left to learn. We studied the eyes and skin and bones and facial cavities and systemic autoimmune diseases all at once. But it also signaled the final push. We had passed the final examinations meant to test our knowledge of the major organs within the body. We were getting close.

During this final unit, time was also "officially" spent preparing for the Step 1 exam, whether that meant reviewing old topics or preparing resources and a study schedule. Over spring "break" the university provided a first attempt at a real practice Step 1 test ... a baptism by fire, and one that "encourages" students to "prepare a bit more" before being baptized again. In other words, many of us didn't do well.

A conflicted energy swirled out with us as we opened the doors after the last final exam of our second year. The fifth unit was over and we were done with the classroom portion of our medical education. It was both a satisfying and frightening realization. We were left with between four and eight weeks to review the information we learned in the first two years of medical school. All of the information. It was a grind, but sitting at a desk in isolation memorizing facts was something we had all become good at.

Questions on the Step 1 are in vignette form and often are written in such a way as to make a test-taker have to know several layers of detail to get to an answer. Instead of asking, "What is this disease?" they ask, "Based on this image of a biopsy from this organ, what side effects would you need to consider when administering the proper treatment?" This is much closer to the thought required of a physician in the actual clinic than a one-liner followed by single word answers. Instead of passively reading text or recalling individual facts with flashcards, practicing with well-written questions coaxes the brain into navigating old connections and stumbling across new ones in the process. That said, it can also be unbelievably frustrating to understand the first two or three levels of a question, only to forget a final piece that will get you to an answer. Seems to me there should be some partial credit on those ones.

Another element of difficulty they've included is that the test doesn't have different portions categorized by organ systems or themes, so anything is fair game at any time. A question about an old man with chest pain is followed by a question about a young girl with a sore throat, which is followed by a question about the steps in a metabolic pathway. The mind must stay wide open, as well as able to follow even the trickle of an idea to its microscopic conclusion.

During the final week of preparation, recognition and understanding were at all-time highs, as were material-load and urgency. All that was left was to try and maintain what had been learned the previous five weeks. It was a delicate balancing act, trying to add a few more bricks on top while making sure the base didn't start to crumble. There was also an element of gambling: What will be on my test specifically? Should I make sure I have the side effects of all of these antipsychotics down pat? Or would that time and memory space be better used cramming in these lysosomal storage disorders? There was no right answer, and in the end, as long as you kept studying and stayed relatively healthy, it was a success.

The day before my exam, I had only one cup of coffee, which was perhaps a quarter of my daily intake toward the end there. I exercised, studied until 5 p.m. or so, and closed my First Aid. A somber last supper with my girlfriend followed. As I bid farewell, a single tear traced a fresh path down my ink-stained cheek. All would be over soon.

Game Day

I felt good the morning of. Somehow, I had stayed focused and on-schedule during my study period. The practice tests I took hinted that I was prepared to do as well as I hoped on the real thing. I was relatively well rested, had packed sufficient calories and caffeine to get me through the eight hours of exam, my lucky rabbit's foot was around my neck, and my lucky socks were on my feet. Nothing was going to faze me.

Right away I had to take off the rabbit's foot because the chain it was on set off the facility's metal detector. Not ideal, but I still had my socks, and the power of a rabbit's foot can certainly travel through the thin aluminum of a locker, around a corner and under the testing door. I was still at nearly full strength. After an ID check, a few fingerprints and a signature, I was sitting at the computer, staring at the first of over 300 questions.

My second question was one that included audible heart sounds. Unfortunately, those sounds weren't coming through my headphones. This was the one type of question I wasn't prepared for: How do I stay calm through a software malfunction? After a moment of troubleshooting, which included looking around frantically, replugging in my headphones, and toggling the volume up and down, I raised my hand. The test administrator promptly helped shut down my station and relocate me to a nearby desk, which happened to be the one labeled number 15, my lucky number. That was a very good sign at a crucial moment.

I lost almost two minutes during the fracas and was rushed toward the end of that first section. But I figured there were going to be much more difficult problems to work through as a physician, and that perhaps an audio malfunction was just the part of the exam meant to test patience and resolve under pressure. Well, that's what I told myself so that I didn't punch something. The rest of the test was relatively unremarkable, aside from its length and difficulty.

The Aftermath

I feel more positively about the Step 1 experience than I thought I would. Back in the winter days of second year, I knew the exam was something I had to get through and that it was very difficult for a variety of reasons. But I didn't know then what it would actually look like for me to organize a giant body of information and systematically wrap my arms around it. Moving forward, I am more prepared and confident in my ability to build something large using small steps. It remains to be seen how much of the material I will continue to use regularly and how much will be repackaged as a vague intuition I rely upon when out of conscious ideas. But for the time being, it feels important to have solidified the concepts and essential facts from the first two years of my training, even if the strange names of all of the drugs I just memorized change before I can legally prescribe them.

Now it is time to let out the stale breath I took in long ago, take apart the makeshift standing desk I constructed out of cardboard boxes, burn a few piles of notes, and get ready to interact with the real world. The time for studying the principles behind the actions is finally over. Now it is time to see what health care providers actually do.

The Battle of Midday

December 1, 2015

Valerie Efros
Michigan State University College of Human Medicine
Class of 2018

I WROTE THE FOLLOWING POEM about two weeks before my USMLE Step 1 exam. By this time, I had been in my intensive study period for nine or so weeks, and my faith and confidence were fading fast. Despite my best efforts, my struggle to digest all of First Aid for the USMLE Step 1 continued, forcing me to push back my exam date. I was stuck in this never-ending study bubble, while all of my friends continued onwards with excitement for their first clinical rotations. Studying had been so isolating day in and day out, and the terror of the disappearing days was sinking in. Every day was a battle not only with the endless facts I had yet to fully memorize, but also with my own spirit, willing it not to be broken. Yet in this moment of self-expression, I found myself sprawled out on my bed, red notebook in hand, feeling inspired by how far I had come and how hard I had worked. The pride and determination that flowed through me was both refreshing and unexpected. I did not think; I just felt.

This poem is a culmination of the exhaustion, perseverance, willpower, fear and sacrifice that represents the benchmark in our journey through medical education. I hope the feelings I convey resonate with others who have gone through the same, for these are the strings that connect us all, regardless of university or hospital affiliation.

—

Terrified beyond belief I never thought I'd be.
I'm overcome, I'm overrun, with little left to show.
I'm standing still, the world a thrill, my watching from within —
this room of mine, it's time to shine —
to show them I'll succeed.

It's never been an easy road, the one I choose to go.
Sometimes regret, with tears and fret, I let the demons know;
that I'll be strong, yes, all along, I'll fight with blood and sweat.
For to this day, I still can stay, that I've returned —
though badly burned —
from every battle I have met.

Not won or lost, but worth the cost, of standing up to fear.
For I'm the one who fought the sun;
I'm standing. I'm still here.

And so it flows, the story goes, just as it did before.
With mind unwound and silent sounds,
I've tallied up the score.
Though not so sure I made it clear
It has not been with ease.
I'm off the grid, it's time to live
The battle of midday.

Escaping the Sphere

December 2, 2014

Diem Vu
Mayo Clinic School of Medicine
Class of 2016

STEP 1 STUDYING CAN be a lonely endeavor. This is true even if you have a study buddy — you may share a table at the local café, but you might as well be sitting in two different worlds. One of you reviews cardiac physiology while the other watches a renal pathology lecture. Your worlds may convene occasionally as one of you quizzes the other on neurocutaneous syndromes or shares an interesting tidbit from a question bank ("Did you know that Clostridium botulinum lives in Californian soil? I didn't ... oh NO!").

So unless you carefully coordinate your studying with a friend, the bulk of your time is spent within an invisible sphere the radius of which is determined by the distance between you and your caffeinated beverage or the closest power outlet. Its boundaries are clear, but impenetrable, buzzing like an electric fence — just loudly enough for you to remember it's there.

This magical hamster ball doesn't have to be completely sinister. It can be a protective barrier from distraction and procrastination — a sign to any friends who approach you that, no, this is not a good time to rant about something offensive Iggy Azalea tweeted or about their apartment's maintenance guy's hallway shenanigans.

However, the best advice I can give anyone studying for Step 1 is to try your best to occasionally poke your head outside of the sphere: Take some great big breaths of fresh air, listen to the rain pattering on the window, enjoy the warmth of sunlight on your face.

Even better, you can leave your sphere propped up in the corner of your apartment as you pick up your guitar and sing for a while. Or you can leave the sphere behind you when you go running outdoors and listen to your current pop guilty pleasure. Roll it into the yard and cover it with a blanket, then spend time with your significant other, friends, parents or children. Leave it in the car when you go to the mall, movie theater or restaurant.

The trick to abandoning your sphere is to not feel guilty about it. This was a struggle for me. With such a huge depth and breadth of information tested on Step 1, how could you possibly justify goofing around for a few hours? Plus, you'll hear rumors that some other kid in your class studies for 16 hours a day and hasn't left their desk to sleep or eat in two weeks. (Also, they're a cyborg).

But stepping out of the sphere to talk to loved ones (who weren't studying for Step 1) reminded me that if I stepped away from my First Aid book for two hours then the sky would not split in two and belch a rain of apocalyptic magma. Life would move on, and the future was not determined solely by one three-digit number.

And after all, you can't swim the length of a pool without taking any breaths. The same goes for living in the invisible sphere.

Eat, Study, Love:
A Guide to Surviving the Boards

January 18, 2013

Valentina Bonev
University of California Irvine School of Medicine
Class of 2014

STUDYING FOR THE BOARDS is like preparing for a marathon shelf. Your stamina, knowledge, guessing skills and sanity will all be tested, although these are not formal topics listed on the syllabus.

At first, you're gung-ho and ready to crack open your freshly bought books. Then, you slow down as you stare out the window wanting to be outside enjoying life. Lastly, you stare at your calendar wishing your test date was already here so you can get it over with, or you panic and reschedule your test date.

So what do you need to do to make this process go smoothly? There are some simple things you can do right now. First, drop that bag of greasy potato chips and stop drinking that soda. It's important to eat healthy, balanced meals. If your stomach is happy, so is your brain. Don't be afraid to eat comfort foods that make you happy, as long as you don't splurge too often. Some people will make a delicious meal or go out to their favorite restaurant as a way of rewarding themselves.

You need to make time to blow off some steam and keep your body sound. Make time to play basketball or whatever sport you like to play. Unfortunately, most of your friends will be studying, so you'll probably be by yourself.

Now it's time to actually start studying. Different methods will work for different subjects. An outline may work for anatomy, but flashcards may be best for microbiology and pharmacology. It may help to form a study group where you teach each other topics once or twice a week; play off each other's strengths and weaknesses. Don't get nervous, however, when one colleague is studying pathology while another is on epidemiology but you are studying physiology and haven't even begun those two subjects. Don't worry, because everyone is on their own pace. You may not even study biochemistry, for instance, if you majored in biochemistry during college. So, keep stress to a minimum and stay away from people and situations that stress you out.

Set goals and constantly reassess yourself. It's okay if not everything goes as planned, just adjust accordingly. Reward yourself when you meet goals. Also, stay organized: this will keep you on track.

Unfortunately, you won't be able to make every wedding, get-together, date night and video game night because you'll be busy studying. These are just small sacrifices that we have to make to meet our long-term goal. Friends may pester you and they may not understand why you spend so much time studying, but true friends will stick by you.

You may get to the point when you're sick and tired of studying. Before you pull out your hair, remember that this information will help you with your future patients. How so, you ask? For example, if you're a psychiatrist with a patient with an abdominal aortic aneurysm (AAA), you must be able to recognize the signs of a ruptured AAA in case this happens during one of your sessions, because it's an emergency. Maybe you become a pediatrician and you use embryology to explain to the mother of your patient about her child's Tetralogy of Fallot.

Do not be discouraged by what seems like failure; you cannot know how sweet success is until you have failed. One reproductive endocrinologist I worked with put a positive spin on our successes and failures: "Life is about victories and learning experiences."

Grandmother

February 21, 2018

Aleksandra Bacewicz
University of South Florida Morsani College of Medicine
Class of 2018

I DID NOT KNOW I was feeling sadness until I found it hard to swallow. *There is no reason for it*, I thought. At 94, she is still sharp, most of the time. The closer my Babcia, or grandmother, is to death, the more irreverently she talks about politics. She is progressive and possibly radical for this town of barely 60,000 traditional (read: Polish Catholic and conservative) citizens. She has two sons who are accomplished in their respective ways; the younger of the two is my dad, or my Tatuś, as I say endearingly in Polish.

There lies a mass of asymmetrical triangles in the spectrum of amber and crimson. Her entire body is covered with only her head peeking out from the heaviness of the duvet. The material looks like it is from the 70s, clashing with the faded limes on my grandmother's peach-colored top. Beams of an autumnal sun rest upon her cheeks until they are jilted when she hears the door shut.

"Ola, Ola, to ty?"

She uses my Polish name, short for the extravagantly Slavic Aleksandra, asking if it is me. I do not respond, I can't. Her home health aide greets me with an emphatic *yes* and my grandmother's eyes reach beyond the bed, beyond the doorway, straining to find me.

I sit at her bedside. I am within a foot of her, a twitching as her fingers clasp and unclasp mine. She looks toward me but her gaze travels on the periphery of my frame. Her countenance is lit from within, a piercing awareness outweighs any sense of loss in her dulled eyes. Small fresh lips surround her mouth as she works around the words to extract my story. She wants to know everything: "When are you done with medical school, what do you think of the newest President, when will you move back to Poland, would you please take 200 zloty off my hands, why are you not eating anything, oh and are you in love?"

It would be incorrect to say I am holding her hand at this point. She has pulled me toward her as much as she can from the bed she has now been in for three months. So much so that she has both hands wrapped around my left arm nearly up to my shoulder. I do not mind, as I nuzzle my head into the bedrail.

My grandmother broke her hip over a year ago. On the cusp of 93, it was decided that nothing would be done about it. She continued to move about her home with a walker and started to receive help from informally-trained home health aides. In small towns like these, social services are put together with odds and ends of people's time and skills. She now has 24-hour care split among two aides and a distant relative. The fractured hip continued to cause her discomfort which then extended down through her lower extremity. She became bedridden a few months ago, unable to stand the pain. Although confined to a small space with a weakening body, my grandmother does not complain, occasionally popping an ibuprofen.

Conversation between us coincides with commercial breaks for *Family Court*, the program she flicks on every weekday at 1 p.m. Her afternoon aide shuffles around the room, preparing tea every half hour, chuckling at my grandmother's engagement with the unfolding story on TV. The aide is a retired woman dressed in shades of grey and brown. The only parts of her that grab my attention are her spritely hazel eyes and modest golden crucifix that hangs around her neck.

Time passes quickly. There is no pressure, even in the silence, to say or do anything in particular. She knows my type because she raised my father — quiet, calm, often withdrawn.

I come to see her again the following day and it is now that I feel the weight of every minute. I have a train to leave town that afternoon and I do not have plans to return to Poland for at least the next half year. It is doubtful that she will scrape by for much longer. Although her awareness is intact, she is gradually becoming disconnected from the outside world. She has difficulty keeping up with conversation unless every word is expressed as a shout. Her vision has long been fading and she can only make out dark blobs hovering about her room. She is unable to remember the last time she was outside, as she lives on the second floor of a building with no elevator.

She hears my fiddling with the coat rack, and I slip my shoes off. I see her starting to look around and call my name. The morning aide is here this time. I take in a scent of faint smoke as I pass her, dressed in black leggings splashed with hot pink. My Babcia does not hesitate to pull me close once more. The aide comes over to feed her a *naleśnik z mięsem*, a rolled-up crepe filled with meat and onions. My grandmother takes a small bite and chews until I imagine whatever is in her mouth has completely liquefied. She swallows and stops there, refusing any more food. She turns to lie back on the bed as her eyes flutter and close halfway. I feel like that is appropriate — she seems to be equidistant from life and death at this point.

And this is how it goes, I think. She has lost her independence, for the most part, and her senses are diminishing. Her upper body is still mobile, unlike her lower half. Daytime naps have turned into an early afternoon bedtime. All the signs of an impending death.

The thoughts begin to weigh on me as I am left alone with her slumbering body. Seeping in through the grief is also relief and gratitude — that I am here, that I am able to say goodbye, that she is comfortable at home. I think about the patients I have worked with back in the States who were also on death's doorstep. I barely saw these phases of death as I see them so distinctly now. The aseptic walls and clamor of hospital days towered over life's subtleties and squashed any sense of a peaceful transition. I am glad medical culture in Poland, in general, trends toward being less aggressive, especially near the end of life. It is common for people to die at home and for friends, neighbors and family to care for those who are on that path.

After nudging my grandmother to say a final 'see you later,' I run back at least three times to plaster her cheeks with kisses. I am carrying shards of glass in my chest. I will not be here again, because she will not be either.

"Jestem z Tobą," she says. *I am with you.* And I know she is.

Where Are You From?

June 16, 2017

Syed Shehab
Larner College of Medicine at the University of Vermont
Class of 2017

"WHERE ARE YOU FROM?"
A question that I am asked many times during the course of my day. But the answer has never been clear nor concise.

I was born and raised in Dhaka, Bangladesh. I came to the United States at age 18 for my undergraduate education, and my parents immigrated after my freshman year of college. They settled in Atlanta as we had family there. After college, I lived and worked in New York City and finally moved to Burlington, VT for medical school.

When someone asks the question, "Where are you from?" I find myself weighing these three options: Bangladesh, New York City or Atlanta. But it is hard for me to decide which one to choose.

I feel more American than Bengali. I view the world as a new American. I am ill-informed about the Bangladeshi society, art, politics, media and even the cricket team. A lack of basic cooking skills also means that I rarely indulge in Bengali cuisine — something that is central to Bengali culture. My loved ones all live in the United States and they all have American concerns. This may be the ultimate example of acculturation, appropriation or assimilation, but I now rarely identify with my native land.

New York has been the obvious choice. My formative years were spent there. My personal, emotional and intellectual growth is directly linked to the people, culture and politics of this state. But I have only spent four years in the New York metropolitan area and a little over two years in the city proper. I have no roots in the city. I was very much like the millions of transients who take up residence in exorbitantly priced tiny apartments in search of an authentic and quintessential NYC experience. I feel like a fraud when I claim NYC as my land of origin — as if I was feigning it to look hip and sophisticated.

I have never laid claim to Atlanta. My entire family now lives there year-round but I hesitate to call it home. Having only spent two to three weeks at a time in Atlanta and knowing only half of the tourist attractions in the city, I remain oblivious to what makes the city and its people tick.

As uneasy as this convoluted sense of belonging makes me feel, what truly puts me ill at ease is the need a lot of people place on knowing this information.

There is the innocuous "Where are you from?" and then the probing "Where are you from?" The person who asks the latter question seems to always know the answer, always expecting me to conform to their image of who I am. The nurse who doesn't believe I am from New York City — "You don't sound like you are from New York!" The technician who is incredulous that I am from Atlanta — "You are not a southern boy." And the doctor who seems perplexed when I state that I am from Bangladesh — "I thought you were Indian!"

But it is always difficult to tease apart which "Where are you from?" question is being asked. As I thought about the distinction between the two, I came up with what I thought would be the perfect answer. It was a pithy answer, similar to the answers everyone else would give. It wasn't the whole truth but it wasn't a lie either. And it kept me from having to continuously retell and explain my immigrant experience to strangers. It also allowed me to judge which of the two questions was being asked.

Q: "Where are you from?"
A: "I am *recently* from New York."

I have routinely used this tactic as I changed services and hospitals regularly as a third-year medical student. This approach is not 100 percent effective. But it has been surprising to note the number of people who have very similar replies to my answer.

Q: "But, where are you really from?" Or,
Q: "Where were you from before New York?" — As if it is hard to believe that a South Asian can also be an American.

I have decided that people who have the above reply are asking the probing version of the question.

What I cannot understand is why this information is so important. There is the obvious answer: that it allows people from the same place to form a connection, that it provides a sense of kinship. It can also be used to stereotype people and endow a person with many characteristics and qualities, no matter whether they deserve them or not. It helps people to categorize others into neat little boxes and helps us make sense of the world. And the more nefarious reason is that this information allows someone to feel superior about their standing in life while simultaneously making someone feel inferior. There can be a multitude of other reasons why this question is so important. I am not oblivious to how race, gender, socioeconomic class and perceived citizenship status, among others, plays a role in why this question is asked. Nonetheless, it is astounding how something so mundane is used to codify and summarize people.

Identity issues are not exclusive to medicine. In the current political climate, identity is fast becoming a polarizing topic. I chose to believe that medicine was different. That the intellectual and compassionate nature of this work separated medicine from the typical fallacies of the majority. Somehow the white coat was able to create a true meritocracy. However, we still remain members of a greater society and in that world, race, gender, class and sexual orientation matter. The hospital is not exempt from the daily struggles of America. And this has been an important realization.

If you, the reader, were one of those people using the "Where are you from?" question to establish kinship, to find commonality between strangers — I propose that we use a different question. Maybe,

Q: What is your favorite color?
Q: What kind of books do you read?
Q: What do you like to do in your free time?
Q: Are you a dog or a cat person?

Widening the Discussion of Mental Health in Medical School and Beyond

May 7, 2016

Kathleen Tzan
Sidney Kimmel Medical College
Class of 2017

I WAS UP LATE IN the midst of an intense infectious disease cram session when my phone buzzed alive. Glancing down at the light of my iPhone, I noted the caller's name with surprise — it was a friend across the country with whom I had not spoken with in months. I wondered why he was calling now; maybe it was a butt dial?

"Hello?" I asked questioningly. The voice that responded was calm and polite as usual, but also embarrassed: "Hey, sorry to bother you ... I'm at the hospital, but don't worry, I'm okay. Can I just talk to you for a second?"

My heart sank with the comprehension that I would probably be pulling an all-nighter, but despite the cavalier tone and apologetic chuckles coming through the phone, there was a nervous uneasiness to my friend's voice that caught my attention; this was not a phone call I could turn down. "Yeah sure, of course. What's up?"

—

The Jay I knew in high school had always been a wallflower — someone who ambled through high school with decent grades and a few leadership positions, never all that ambitious, but not for lack of intelligence. It was just his personality to be mild-mannered. In our gigantic class of 800-plus competitive students, Jay blended right in, somewhere in the top 10 percent, but nothing outstanding. We were close friends, but drifted in opposite directions after graduation. I moved out of state, graduated college early and stayed out of state for medical school. Jay attended community college for two years before transferring to a larger state university.

As a transfer student, Jay struggled to fit in. Academics were tough, but it was his social life that really took a hit. Many students had bonded as freshmen in the dorms and had already discovered their own niches in clubs and extracurricular activities. Cliques had formed. People were polite enough, but he was the new kid on the block — on the outside looking in.

Seeking camaraderie and exercise, Jay joined the judo club. The first few club sessions went well; the basic frameworks for new friendships and judo were laid down. Unfortunately, at the third session, the club members teamed up to practice throws. Jay was thrown forcefully over his partner's back, landing awkwardly with his foot smacking the floor off the mat. Subsequent x-rays showed a lateral malleolus fracture, and Jay had to take a break from judo. Budding friendships shriveled up, and he found himself lying in bed with a cast on his left ankle, unable now to even climb the expansive hilly campus to his classes. Assignments and tests passed quickly, and Jay's grades plummeted. At the end of the quarter, Jay was put on academic probation, which threatened not only his status as a provisional transfer student, but also his financial aid.

Pills, a bridge, a gun? No, he wasn't going to shoot himself, but in desperation, Jay found himself spewing suicidal thoughts in his student counselor's office, where he was subsequently given the choice to either commit himself to the hospital's psychiatric ward voluntarily or in restraints; either way, a police escort was involved. Jay described the ride through campus in the police car as surreal. He made small talk with the police officers who apologized for the hassle: *Sorry, kid, for the embarrassment. You aren't allowed to walk to the hospital yourself — strict protocol.* Hours of waiting, paperwork and orientation later, Jay was finally left to himself in the quiet of his own patient room. A couple of days later, he called me.

Jay was discharged on SSRIs, with orders to see a therapist weekly. He returned to school briefly, but the stress of classes, therapy, his still-broken foot, and financial aid probation was too much — Jay decided to drop out of school for a while. He and I stayed in touch through phone calls and met up whenever I was home on break from school. He always seemed so calm and unaltered on the surface, but deeper probing would reveal still gaping holes in confidence and happiness.

In a sense, what happened to Jay could have happened to anyone. There was nothing to be ashamed of, yet I saw how easy it was for friends and family to avoid talking about the taboo subject, act as if nothing happened and move on with their own lives. Even those friends like me, who wanted to be supportive, were busy with our own stressful lives, and it was easy to go months without talking about the issue, which made me wonder.

—

Why is mental health faced with such silence? Why is it that students like Jay are not warned or made adequately aware of the immense toll that academic studies and social pressures can take on the mind, to the point that the first indication of anything out of the ordinary is emergent hospitalization, or even worse, suicide? According to the National Institute of Mental Health, in 2014, about 6.7 percent of adults in the United States experienced at least one major depressive episode in the past year. So why is it that even in the cases of completed suicide today, the overwhelming reaction is still disbelief, as if mental illness is rare?

A fellow student writer recently wrote that she wondered if depression were "just part of life as a medical student." One of her professors had given a lecture on depression asking students to "think of how many people we knew with the signs of depression listed on his lecture slide" — excluding medical students of course, "because you've all got some of these." There is something so terribly and inherently wrong with that statement.

This student writer argued that, "if medical students are just expected to show signs of depression at some point during their education, there's something wrong with the system." And, while I echo that statement whole-heartedly, I would additionally argue that medical student or not, *no one* should ever be expected to show signs of depression for the sake of self-development or career training. No one deserves to fall through the cracks. And when someone like Jay falls victim to mental illness, the situation deserves attention, not a taboo label. We all suffer varying degrees of mental stress from our paths in life, and it is time that we start to talk about it.

At my medical school, there has been a huge push from the student health and wellness department to break the stigma of mental health. All students are allowed three free consultations with a psychologist, and efforts are made to promote mindfulness meditation, a balanced lifestyle and other methods for stress relief. Even so, and even among medical students who should arguably be some of the most aware people on mental health issues, the pressures of academia trump balanced lifestyles, and the stigma of mental illness remains very real. Medical students remain afraid to use certain services for fear of negative repercussions. Who knows what could be written or seen by someone through the electronic medical record? What if a residency director somehow found out about a mental health diagnosis? True or not, horror stories are passed around of admissions committees flat out refusing to review residency applicants' files after finding out about certain medical conditions. The potential repercussions are too risky to seek mental help. And yet, if even we as future health care workers — with all the evidence-based medicine and education to back us up — cannot confront the stigma associated with mental illness and demand the help we need, how can we ever expect our patients to do it?

A fellow classmate suggested that all students be mandated to attend at least one session with a psychiatrist simply to demystify the experience, as well as destigmatize the process for students who want to see a psychiatrist but fear judgment. In my opinion, this may not be a bad idea. I understand that not everyone wants or needs to talk about their struggles, but in an education system where we spend so many hours studying and agonizing over grades, a one-hour break to check in with a caring, nonjudgmental health care professional could do everyone some good.

Society is constantly reminded to eat healthily and exercise to take care of one's body. In comparison, when was the last time you seriously talked to someone about improving their mental health? If social media and the movements on my campus are any indicator, the door to discussing mental health has been opened, but it is impossible to deny that further progress still needs to be made. We all need to be more self-aware of any unhealth-iness we bear in our minds, to reexamine our thoughts and to be honest about our approaches to mental stress. Whether in a group setting or an in-ternal dialogue, it is time to be courageous and widen the discussion — how are you *truly* doing today?

Booster

March 19, 2015

Jennifer Tsai
Warren Alpert Medical School of Brown University
Class of 2019

FOR ME, HEPATITIS B booster shots feel pretty much as pleasant as being sucker punched in the arm. You can imagine that it didn't inspire much elation when I scrolled through my calendar to see, spelled out in big red letters, a reminder for "Hep B #3."

Now, as I reflect, this reminder feels like a victory of sorts.

When you are told during medical school orientation that you must be re-immunized for hepatitis B, you are sent to undergo a three-dose course: zero months, one month, six months. I remember making those appointments while standing next to a shaded window, looking out into the brutish heat of an August in Providence. I remember thinking that six months felt so far away. I remember that earlier that day, I had found an empty classroom and cried with heaving shoulders into my hands. I remember feeling alone, and small, and mistaken.

And I remember wondering, on that fifth day of medical school, if I would make it as far as February. If I would still be a medical student when the time came for the third dose of that blasted hepatitis B booster.

I never understood how I could succinctly answer the question, "How is medical school going?" It is inevitable that people ask, but whenever posed with this question, I always mentally react with sarcastic commentary. "How is medical school going? Well here, let me give you a genuine, concise answer in one sentence and less than 43.7 seconds."

I am continually stuck when posed with this inquiry. It feels disingenuous to use the noncommittal, easy stock answer that's void of any real substance, but the real one is too winding and whiny to offer up casually.

The truth? The transition to medical school has been incredibly hard for me. I am still unsure if this is where I want to be.

I feel separated from the disciplines I was, and am, incredibly passionate about. I no longer get excited for class the way I did when I was in college taking seminars on racial politics, the construction of scientific knowledge, gender fluidity and medical anthropology. I miss reading ideas and synthesizing my own. I feel uninvested in our material — the science of it feels so detached from the politics and realities of the world. I miss books and discussion and critique; I miss critically questioning what is presented as fact. I think our preclinical education often misses the bigger picture, forgets how medical authority operates, leaves little room to work though the impossible questions of ethics and context and humanity. I am frustrated with a curriculum that relies so heavily on the biomedical framework. I often feel shabby and insecure in all sorts of academic and social ways. I constantly feel that dimensions of my individuality are being flattened by the demands of assessments that third and fourth years tell me do not matter anyway.

This curriculum feels displaced from what I pictured a career in medicine would look like. Especially in these preclinical years, our schooling is at once directly tied to becoming a physician, and yet so separated from doctoring. There are also times, scary times, when I realize I have no idea what this career will actually look like as a real profession — that I do not indeed know what I am working towards. People tell me I will have to wait for the wards to know. I worry that when I get there, it will not be what I hoped.

It does not help that surveys show 90 percent of physicians wouldn't recommend this profession for their children.

There is also the new, overwhelming cognizance that there is an infinite amount of knowledge and preparation involved in medical training. There is no end. There is no clear point of termination, and as such, it is up to you to decide where to draw the line of academic sufficiency.

We are told to "be your own guide." This is a reminder meant to provide comfort that your decisions will be right for you, but it is also terrifying. Being your own guide, having autonomy with intention, means simultaneously accepting culpability for anything that goes wrong. It is stressful for me to decide, day in day out, how to optimize my work-life balance. I am young and stupid. I don't love trying to make important decisions about my future while I indeed feel, young and stupid.

Medical school has been incredibly stressful for me because it constantly backs you into a corner and forces you to figure out what kind of student and person you want to be, all very quickly. It is an encroaching force that I have to remember to actively push back against as it invades larger and larger portions of my mind, my time and my space. It demands that you decide what is important — to decide what parts of your being you will allow to go to the wayside when you have an infinite number of facts to memorize, an infinite number of ways you "should be" advancing your career. It asks you to balance personal desire and need with the looming pressure of accepting responsibility for the well-being of others. I think oftentimes, this pressure forces us to strip ourselves of the little luxuries that contribute so much more to our identity than the term "medical student." We embrace this title with arms weighed down by books and expectation, and in doing so, we often lose others. We are runners and singers, readers and writers, painters, dreamers, and significant others. But when there are infinite facts to learn, there is less time to run and sing, fewer hours to read and write, diminished capacity to paint and dream and love.

And then there is also this point of guilt: How dare I sit here and criticize and complain about an incredible opportunity that people quite literally dream about. How do I negotiate the idea of finding difficulty in this experience while remaining humble and grateful for the privilege afforded to me by this institution? It also seems useless to worry about having too little time, when I am told over and over that I will never again have the freedom first year allows.

—

A dear friend recently told me that she thought I was brave for voicing my doubts about medical school openly. This shocked me. I hadn't realized that it was something worth hiding.

Looking back, I suppose I felt early on that there was some shame attached to my doubt. I realize, now, that this notion is toxic, and only served to deepen loneliness and isolation in a place that does not necessarily need to feel lonely and isolated.

Few people are absent doubt in medical school. As I've been more open with my insecurity, I've found more support, more normalcy, and more validation in my beliefs and fears. I've found people who feel the same. These are people I cherish, am indebted to, and care for deeply. I am grateful for the amazing community of peers I have found in my time in medical school — Providence and beyond.

As difficult a transition as matriculating to medical school has been, as much as my doubts are still present, I look back on the fall of 2014 with a lot of joy. Indeed, there were soaring levels of insecurity, creeping levels of worthlessness, and feelings of displacement, but as a whole, my first semester of medical school feels light instead of dark. It is a good place to be, even if I am unsure it is the right place to be.

This is not a well-articulated piece. I have tried to edit sparsely because I don't think this message needs to be sterilized, primped and polished. It certainly has no intention of convincing anyone of anything. It exists purely because I believe validation is one of the most powerful forces in this world.

If you disagree with everything that has been said above, great. I hope this rambly, disjointed reflection has allowed you to feel more situated and confident that medical school is where you are meant to be. If anything I have written resonates with you, spectacular. Know that you are not alone. If you are ever in Providence, come over and drink wine-colored beverages with me on my couch. We can blast Beyonce's "I Was Here" and have some sort of cathartic bonding experience. This is one part of life I will not allow medical school to compromise.

I still wonder what the rest of medical school will hold. What I do know is that when I get punched again by that hepatitis B booster, I will have at least one reason to smile.

I made it to six months.

Cheers to that.

When students talk to doctors who were in medical school a few decades ago, they frequently remark, "Back then, we just couldn't see how we could possibly memorize all of the information in our books." The medical student today often wishes that he or she could have trained during that time. Now, our "review" books that cover the "essentials" of a given topic are often hundreds of pages long. Given the exponential increase in clinical and research information (ever-growing guidelines, research advances — over 2,500,000 new research articles are published each year), a never-ending influx of new study resources, and ever-expanding medical school requirements and offerings, it's no surprise that medical students encounter academic challenges.

As you see in the following representative pieces, this deluge of new information can lead to stress, professional uncertainty, isolation, and burnout, all commonplace buzzwords in medicine. Medical students should be encouraged to develop the self-awareness and maturity to needed to handle these issues and reach out to others to find support. We hope that, eventually, educators identify a clear means to streamline this academic information rather than simply continuing to add to the pre-existing facts and concepts.

Academic Challenges

OCD by Leor Arbel
Class of 2018 at University of Central Florida College of Medicine

Learning To Be Mediocre

April 26, 2016

Daniella C. Sisniega
Boston University School of Medicine
Class of 2018

M EDICAL SCHOOL IS A constant, never-ending cycle between success and failure — sometimes one occurring within moments of the other. To be a medical student is to fail. We fail at the small things: working out three times a week, being on time for a friend's birthday dinner, working on the research that has been on our desk for months. We also fail at the big things like exams, practical skills, asking for help when we most need it and sometimes letting ourselves sulk for too long.

The second week of my first year of medical school, several other students and I were at the hospital for our Introduction to Clinical Medicine class. We were conducting the first patient interviews of our lives. Two out of the six students in our group would interview one patient each week and then receive feedback from our fourth-year mentors. As you probably guessed, I was terrified. *I'm going to make a fool of myself*, I thought. It was funny and maybe even a little bit hypocritical, how early clinical exposure was one of the things that had attracted me to Boston University School of Medicine, but yet here I was in my oversized, ill-fitting white coat crossing my fingers and toes every week hoping that it wouldn't be my turn. Somehow, I managed to be the second to last student to interview a patient, which had turned into a personal victory just by itself. After watching four of my groupmates interview their patients, I had taken mental notes from the feedback they received and thought I was as ready as I would ever be. I knew that whatever was to happen inside that patient room would not have any consequence — this was just for practice, it was just a time-to-start-feeling-like-a-doctor kind of thing. Yet, as soon as it was my turn, I could feel myself shaking. *Just do it, Daniella, it will be fine*, I said to myself as I walked in, introduced myself and my group and started the interview:

"Can you tell me what brought you to the hospital?"

The patient mentioned an exacerbation of some disease I had never heard about before.

"Okay, um, did you have any pain?"

The patient points to her chest. Yes! Great! At least I know the heart is located in the chest. She took some medicine that I don't even know how to spell, so I just scribble something to look engaged.

"Can ... you duuhhrs...scribe the p-pain?"

Oh no. Oh no. The room is spinning. My head is foggy. My mouth is dry.

"E-excuse me. I-I'll be right back."

I held on to the wall and barely made it outside of the patient room before I fell to the ground. Yes, I, the proverbial medical student my parents were so proud of, had fainted on my first patient interview. A nurse walked towards me and asked if I was okay or needed some water and then just yawned and went on with her day. Clearly, I was not the first student to faint. Then, our fourth-year mentors and the rest of my group came out and started saying how it was okay to be nervous and that I would do better the next time. *Yeah right, just sign me up for pathology right now*, I thought while I tried my hardest to smile and act as if everything was okay. I had managed to convince myself on the first day that I was one of those students that was just "bad with patients."

I knew there was no way I could improve other than interviewing more patients. Lucky for me, the chance came during the second semester of first year, when we were all sent to work side by side with a physician once a week. Every week, Dr. B gave me a list of patients to interview and I would meet her later to present the patients and do a physical exam. *Gulp*. It wasn't always great and it wasn't always pretty, but it was always a little bit better and that was good enough for me. It took me several weeks to snap out of it and find that sweet spot where I was confident with my skills, but I did it.

If I learned anything from my first two years of medical school, it's that failing and being mediocre is a bit of a skill itself. It's almost as if you need to practice failing. You are going to fail many, *many* times at many, *many* things. Presenting your first patient? That's a tough one. Your first write-up? Likely to be a disaster. Writing your first abstract? Brace yourself for draft #15. Contrary to the way most of us in medicine have lived our lives, the point is not to go through medical school without failing. It is okay to fail and it is okay to be mediocre when you try something new for the first time. The point is to make failure part of your medical student experience, so that eventually, you will learn enough from it that you will do it right when it matters. It is okay to have fear as long as it does not paralyze you, just as it is okay to fail as long as it doesn't mean you give up. Give yourself a little bit of an opportunity to fail. Trust me, just start, just do it. You'll be mediocre for a while, but it will not be forever.

Are We All Impostors?

October 30, 2017

Alayna Sterchele
University of Central Florida College of Medicine
Class of 2021

Is it just me?
Or does it seem
that my pride
will not let me be
vulnerable
in front of
you.

You must not see
how it takes me
days
to understand the same
concepts
that take you
hours.

That my own insecurities
are hidden behind a
facade of smiles,
a demeanor of confidence.
The first part of act one,
but you are
my only audience.
My grand performance ensues.

Usually I am an open book,
words spewing off
crowded pages.

But around you,
my book is cast shut,
locked away, no title.
Hidden in the back of my mind,
as laughs and easy conversation
deceive you.

I don't want you to see
my weaknesses,
so they stay hidden,
like wounds concealed
beneath faint scars,
only noticeable
if you look closely.

Why do I hear
"You are not the only one,"
yet I am performing in a
one-man show.
Surrounded by actors
but an empty stage.
Everyone is honest,
although, no one truly is.

My struggle is a weakness
that marks me with a
Scarlet Letter,
so that I am no longer
your equal.
I am tainted, contaminated,
like I am generic but you are
the trade name that only
the best can afford.

So I continue to shield myself
behind my armor of
forced smiles and
easy conversation,
in the castle of
my mind
and the fortress of
my body,
that can crumble only if the
cracks are seen, until the day
these walls are no longer needed —
a day that may never come.

It's Hard Keeping a White Coat Clean

October 27, 2014

Claire McDaniel
Georgetown University School of Medicine
Class of 2019

A S I WAS STANDING in my apartment building's laundry room scrubbing away at a stubborn coffee stain, I kept up a steady stream of curses at my white coat. In the seven weeks since I'd first donned it, my coat had apparently decided that it preferred to be any color but white. A Tide-to-Go pen is now a permanent fixture in my pocket, and it's used almost as often as the actual pens.

It's odd how much can be invested in a single article of clothing. The white coat is supposed to be a testament to clinical respect and cleanliness, coffee stains not withstanding. It acts as the uniform of our profession, an unofficial signal to other physicians and medical students that we are kith and kin. It certainly helps perpetuate our own beliefs that we are the white knights, riding into battle against disease and suffering.

Somehow, though, I feel like an impostor when I wear it. I'm a first-year medical student, besieged by biochemical cascades. What do I know of the responsibilities of truly wearing the white coat?

Every time I have put on my coat and gone into the hospital, I have been asked for directions, been given priority for getting on elevators or other small acts of respect. Frankly, I don't feel like I've earned that respect. Not yet, anyways.

First-year medical students like myself have barely dipped our toes in the waters of medicine. Attendings ask me what tests I would like to order for a theoretical patient, and all I can do is stare blankly at them and mumble something I heard once on an episode of "E.R." We simply don't know enough to be able to answer most medical questions. If you want to know about adrenal cortex hormones or diabetes, though, I'm your gal.

I might not know a lot yet, and studying every day is exhausting. But when I trudge home from the library after learning an endless series of enzymes that all manage to sound the same, I see my white coat in my closet. It's a symbol of perseverance, a reminder that I — and my fellow students — took an oath to which we must adhere for the rest of our lives.

The coat is a symbol of responsibility, binding us into the roles we swore to fulfill: healer, advocate and student. Perhaps we don't know enough to treat patients (spoiler: we don't), but the coat is a promise that we will, a promise that we will strive to deserve the respect with which it invests us.

Standing on stage at my white coat ceremony, I wasn't aware of that promise. I was too giddy to think that far ahead. Along with the thoughts of "Can my parents see me?" and "Don't trip!" that were running through my mind, there was still a sense that something larger was being bestowed, something more important than a boxy white cotton coat.

We were told when we first put on our coats that we were entering the medical profession. They just didn't mention that, for better or for worse, we were being pushed in the deep end and told to swim. Don't get me wrong, I am elated at the trust that is being placed in us. However, it places the onus on us to live up to that respect. And that is the true power of the white coat.

Learning The Textbook Case

November 1, 2017

Kaitlyn Dykes
Sidney Kimmel Medical College
Class of 2019

STARING AT EACH high-yield line in First Aid, attempting to commit every word to memory, hour upon hour, is the life of a medical student. The stress, isolation and over-caffeination, amidst the constant influx of information, is overwhelming and can cause even the most compassionate student to forget why they are studying.

At first, it may have been to learn medicine and then apply that knowledge to positively impact others' lives. Unfortunately, it is not uncommon to inadvertently view studying as a resented duty, dehumanized and far removed from patient care. However, this outlook is erroneous. Statistics are the condensed stories of individual lives. Each textbook case is somebody's loved one from whom we have been afforded the opportunity to learn. The information from a textbook case — the classic symptoms, physical findings and disease sequelae — is highly privileged and life-saving.

In the library, feeling beyond fatigued, I came upon a list of risk factors for the development of deep vein thrombosis. My breath caught as a chill ran down my spine. My cousin died after a long car drive when she was 18 due to a pulmonary embolism caused by a deep vein thrombosis that had traveled to her lungs. She had all the risk factors identified on this list and yet was prescribed medication, which she was likely contraindicated for, less than a month prior to her death. I cannot help but wonder perhaps if those providing her care had recalled the list of risk factors in front of me. Had they applied it to my cousin's case, could she have had a different outcome?

A few weeks later, I came across the classic description of renal cell carcinoma. Within that condensed paragraph, I was amazed to find the final years of another loved one's life described. The high prevalence of recurrence — news that rocked my family — was printed clearly as an expected aspect of this disease. It dawned on me that, by learning textbook cases, medical training yields the ability to predict upcoming hardships of patients; along with this, there is the responsibility to guide patients and their families in navigating these hardships. What is not in the textbook is the perseverance, strength and love brought out by diseases. Further undocumented is the pain and suffering illness inflicts. Ultimately, the humanity of medicine remains unwritten amongst the high-yield facts and thus becomes the responsibility of the student to write in the margins as they advance in training.

Recently, I found myself acutely confronting this challenge as I walked down the halls of an unfamiliar hospital, looking for my first real patient's room. This was the first time I was going to take a history and perform a physical alone — and I was nervous. Finally I found the room, took a deep breath, re-adjusted the strange clanking medical tools in the over-filled pockets of my short white coat, rubbed in the hand sanitizer and knocked on the door.

I began with a simple open-ended question and then listened to my patient talk. By the conclusion of the history and the admittedly bumbled physical exam, I felt I had learned a great deal about this patient and his life. I was surprised by how closely my patients' disease mirrored what I had learned in class. I was also proud that, based on my patient's account of "sky-high sugars" and classic symptoms of increased thirst and urination, accompanied by confusion upon admittance, I was able to deduce that he likely experienced a dangerous complication of type II diabetes: hyperglycemic hyperosmolar syndrome.

Later, I presented the case, sounding eerily like the automaton echo of the question stems that are cornerstone of exams, "Patient is a 78-year-old male..." At the conclusion of my report, the attending pointedly inquired, "What caused the patient to reach this point of uncontrolled hyperglycemia?" The attending was not referring to the pathophysiology that causes hyperglycemia but instead to the fallible life circumstances that unfolded into the patient's state of deteriorated health. I did not know the answer to this question. I had not learned the most important information pertaining to my patient's care at this point in time. Did an infection tip this patient's homeostasis? Was there a breakdown in his social support system? Was the patient non-compliant due to lack of understanding, economic impediments or depression? The answer to the attending's question was pivotal in understanding the patient, his current illness and identifying the most efficacious way to prevent this life-threatening state from reoccurring. I had failed to recognize how intrinsically human aspects of this disease amalgamate with its textbook description, and as a result I did not fully understand my patient's story or how to best care for him as an individual.

This experience was a wake-up call, a true reiteration that medicine is more than the study of disease, but also the practice of discerning human nature. Therefore, it is the medical students' responsibility to read the subtext of humanity amongst the high-yield facts. By learning as much as possible and unifying pathology with the person, students can become adept physicians able to apply this knowledge — not only to accurately answer multiple-choice questions, but to treat each patient uniquely and therefore correctly.

The Burden of Knowledge

March 30, 2017

Kshama Bhyravabhotla
Morehouse School of Medicine
Class of 2018

I'VE HEARD IT SAID that knowledge is power, and that to be forewarned is to be forearmed. I still remember getting a text from my mother when I was on my OB/GYN rotation, during the first window of time I had gotten to use the bathroom all day. I remember her texting me a picture of a CT scan of my grandfather's lungs with the words: "What does this mean?" And just like that, my grandfather was reduced to a standardized test question stem: *An 80-year-old man with a 30 pack-year smoking history presenting with a spiculated lung mass on CT.*

Everything ceased to exist except for the picture on my phone, the chill stealing its way through my veins and a vague buzzing in my ears. I tried to warm my blood with a few fortifying breaths that caught in my chest and didn't seem to be making their way into my lungs. I blotted my suddenly clammy hands on my scrubs and tried to go back to work as if nothing had happened.

My family in India is usually daunted by crisis and plagued by indecision. Once the doctors in India confirmed the diagnosis of lung cancer, things progressed at the speed of light. My grandfather, a typically proud man who made decisions with his heart and did what he felt was right, told his children to make whatever medical decision they thought was best. He was scheduled for a lobectomy. My mother immediately flew to India to help care for him.

I've heard it said that knowledge is power. But sometimes, knowledge is the torture of knowing exactly what can happen and realizing that you are powerless to change the outcome. I couldn't help but think of the many hours I had spent over the past two years studying the different types of lung cancers: how to treat them, which type caused hypercalcemia, which type caused hypercortisolism. How tawdry this paltry litany of facts felt now, useful only for blind regurgitation onto test papers, as my grandfather lay in a hospital bed on the other side of the world. To hold on to any semblance of control of the situation, I called my mother every day, sometimes multiple times a day. Beyond the few developments she was able to give me, we mostly spent our time on the phone in empty reassurances and silence, because that was preferable to hanging up and being left alone with my own thoughts.

Call it faith or desperation, but the heart's ability to cling to any available shred of hope is infinitely more powerful than any knowledge obtained from books. I always thought I would be better equipped to handle death after spending two years in a classroom learning primarily how to intellectualize illness. But when my mother called me that afternoon to tell me that my grandfather had passed away due to post-surgical complications, I sat frozen and uncomprehending on my couch, feeling no less lost than the first time I had encountered death. Until that moment, I hadn't realized just how much I had invested in the slim chance that he would make it. Never mind everything I had learned about respiratory distress; never mind the fact that every time I visited him, I had been watching the laughter lines fade from his face and the light dwindle from his eyes over the past ten years, replaced by a desperate desire to be somewhere else. All I could wrap my brain around was the fact that my grandfather, the cornerstone of my spirituality — the impossibly tall man who would steal my toys from me when I was a child and wave them merrily out of my reach until I was exasperated to tears — was not only fallible but gone forever.

As physicians, the burden of knowledge is one that is not only placed on us but one that we shoulder with eagerness. We are bred in a culture where it is imperative to know everything and to fix everything. Understanding this fact, it is even more important to recognize that our knowledge does not replace our innate need for humanism and compassion. Our relationship with death is perplexing; it is a constant see-saw between being paralyzed by our fear of failure to the point of forgetting our patients' emotions and numbing ourselves by accepting that death and old age come hand-in-hand. Ultimately, our knowledge does not protect us from the ironclad truth that we are not infallible and that, even after we study our hardest, an element of the outcome is beyond us. The surgeon who performed my grandfather's surgery was so shaken by his death. It reminded me that, while doing less than our best is not an option, our illusion of having total control is fragile and evanescent.

My grandfather's death also challenged me to look at patients in a different way. He was a deeply spiritual man, more concerned with the soul and transcendence than the physical body and materialistic trappings of a world for which he no longer cared. He always told me that he hated seeing his doctors because he never felt respected as a person. It always brings a pang of pain to my heart to think of how much he must have resented the fact that the end of his life was marked by the final ignominy of visiting doctors who reduced him to a bundle of flesh and fascia, muscle and matter, with their poking and prodding. Remembering how his personality seemed to vanish piece by piece before his death, I challenged myself to see elderly patients for the people they were before they came to the hospital — for their stubbornness, their curiosity, their fierce independence, their aspirations. I resolved to remember that all of them had once been filled with the same passion and dewy-eyed excitement for life's possibilities that I have today. No one feels the pain of losing that vitality as acutely as the patient.

Whenever my grandfather and I talked on the phone, he would always make fun of me and call me a "half-baked doctor" because I hadn't graduated yet. I know that he was proud of me before he died. With his blessing, I hope to spend the rest of my life in the knowledge that I am working in the shadow of something bigger than myself, and I hope to always treat my patients with knowledge, humility and compassion ... in other words, to become a full-baked doctor.

On Fear, Failure, and the Future: What Med School Can't Teach You

October 12, 2015

Jennifer Hong
Emory School of Medicine
Class of 2018

AS I SETTLE INTO MY second year of medical school, I'm confronted with the fact that I'm one-fourth of the way to an M.D. — that an entire year has passed, and unsurprisingly, all those predictions my deans made at the very beginning came to pass: time flew, we learned more than we thought we ever could, and upon close self-examination, we're very different from how we were this time last year. I understand bits and pieces of medical jargon that, quite literally, can be a completely different language. When I pass a CVS, I think of "cardiovascular system" instead of the pharmacy. CT no longer means Connecticut; it only means the scan. And why do I have to explain what nephrology is to other people? Nephrons are clearly in the kidneys — it's such a common term in daily lingo, after all.

The funniest thing about medical school is that despite all the studying and all the learning, I am constantly in a state of anxiety of how much I don't know, or fear of how much I am capable of forgetting. There is always the looming, ominous presence of Step 1 lingering, followed by the fascination and fear of the wards, and of course the prospect of matching into residency as well.

At every step of my academic career, I have been nervous at some point or another. Being concerned is a natural part of the process — I believe worrying about my academic performance is partially responsible for where I am now. But being afraid academically in medical school has been an entirely different experience from before. For the first time, I am not concerned about excelling — I'm more afraid of legitimately failing, and this fear became clearest when I failed an anatomy exam. For once, this "fail" was not hyperbolic; it was a legitimate delineation that I had failed to achieve an academic baseline that I'd always expected not simply to meet, but to exceed.

In the same time span of the last year, as I was confronting very real external and internal challenges to my own intelligence, the social landscape of the United States was changing dramatically.

On August 9, 2014, unarmed black teenager Michael Brown was shot by a white police officer. This caused the re-launch of the Black Lives Matter movement, one that has become increasingly relevant as what seems to be an endless stream of cases of violence against unarmed black individuals reach national headlines: Tamir Rice, Eric Garner, Walter Scott, Freddie Gray, Eric Harris, the Charleston Emanuel African Methodist Episcopal Church shooting, Sandra Bland and many, many more. Contrary to what it may seem like, the Black Lives Matter movement and all others associated with it did not spring up overnight. Racial profiling and mass incarceration of black and brown minorities have roots in the fairly recent War on Drugs that was declared by Reagan and further exploited by Bush Sr. and Clinton. Essentially, the social landscape of the United States did not so much change as it shifted decades of simmering tensions into national spotlight; social media and an entire generation of tired youth forced these issues to finally require everyone, not just those directly involved in those communities, to examine the honestly deplorable state of racial and class relations in the United States.

In the last year, two other incidents directly concerning the medical field also came to light. The first was the release of the U.S. Senate investigation findings of the CIA interrogation tactics in December 2014. The findings showed that the CIA's torture techniques were not only ineffective, but were also deeply misrepresented to the public and its policymakers. Additionally, the CIA paid two psychologists $81 million to develop the "enhanced interrogation" program that included a host of techniques, of which waterboarding was the barest minimum. In response to this news, Harvard Medical School professor and journalist Dr. Atul Gawande wrote a series of tweets condemning not simply the CIA torture report, but more specifically "how doctors, psychologists, and others sworn to aid human beings made the torture possible ... doctors were long the medical conscience of the military. The worst occurred because government medical leaders abdicated that role."

In a similar but less well-known vein, in June 2015, 243 doctors, nurses and other medical professionals were arrested in a national Medicare fraud that amounted to over $700 million in false billings of equipment or medications that were not needed. Many of these patients were also mentally ill and were not capable of fully understanding their situation or giving consent.

In our second-year orientation, our deans brought up the very serious topic of mental health among medical students, especially in this year where we will start studying for Step. We're told that it is common to start questioning our choice to enter medical school and our capabilities to succeed here. We are asked to watch out for each other, to recognize any changes in our friends' patterns that could signify depression.

And it's easy to hear those words, to agree with them. It's another thing entirely to really process how they can apply to you personally. Because the thing medical school truly can't teach you is how to really deal with fear and failure in addition to the loss of faith in the profession you want to believe is supposed to be the epitome of simply "helping people."

We belong to the profession that swears to do no harm. But what do you do when the medical community remains largely silent about the historic social and racial issues that pervade to this day? What about the current discourse about defunding Planned Parenthood; what do you do when you feel like the medical community is slowly but surely getting overtaken by the brutal dichotomy of an increasingly conservative agenda clashing with an increasingly liberal one? What do you do when the medical community itself fails to maintain the Hippocratic Oath? What happens if you see that failure personally — in your attending, on your team, in your community, or even in yourself? Do you speak up each time? Do you repeat every hashtagged name, watch every video, read every article to remind yourself that this is the world you live in, the world you want to change?

I'm going to get personal and be brutally honest. I am tired. I have questioned countless times in the past year if I belong in medical school for multiple reasons. With every new name that reminds the United States of its racial issues, I find myself reconciling the macro with the micro: social change needs to happen, but I also need to pass my exam tomorrow, because if I don't pass, then I don't get to be a doctor. And if I don't get to be a doctor, then I'm not going to be in that position to enact that change. But then I think about how much I've changed in this past year alone and I worry: What if the doctor I become is not the doctor I ever wanted to be? What if I become so jaded that I fail to care, and I become a part of the system that fundamentally needs to change?

These are fears that medical school can't teach away, but in reality, nothing can. "What ifs" are like post-its you can file away if you need to — they're still present but out of sight. I know these are questions that many of my classmates and other medical students have, but despite all of the encouragement our administration gives us to talk about our emotional and mental health in medical school, we still don't want to ask these questions aloud.

To me, med school comes in waves, academically and emotionally. There are moments where I feel like I'm not just above water; I'm in command of a ship, steering effortlessly and excited about the vastness of the unknown. Then that next big wave arrives, and I'm submerged and it's taking everything I can to get out of bed to stay afloat. I think other people are in the water too, but I can't be sure and I'm too stubborn to ask. But when we find others with the same struggle, we can empathize, and through empathy build a platform for community and encouragement, to openly discuss all the weights that are making it so difficult to keep our heads above water. Medical school can't teach us how to not be afraid or be overwhelmed, but maybe we can teach each other exactly how we can just keep swimming.

My Own Little Ice Age

May 5, 2017

Miguel Galán de Juana
Universidad Autónoma de Madrid
Class of 2017

I CAN FINALLY SAY I'M in my last year of medical school. It has been a bumpy ride, but only one clerkship, a research project and an OSCE separate me from graduation. I remember receiving my acceptance letter eight years ago. Thinking back to those days, neither vocation nor sentiment were on my mind. I don't consider myself someone who had a particular passion for medicine. My father is a diplomat and my mother a public servant. The closest my family had come to working in a hospital was my grandfather, who installed oxygen pipes in operating rooms. I stumbled upon medicine, and medicine stumbled upon me. My vocation is something that I've nurtured over the years, watered by the many inspiring people I've met in my journey, and pruned by all the pain and dissatisfaction I vow to avoid inflicting as a physician. I am grateful to have fallen in love with such a noble pursuit. But when I look back to the road that led me here, I feel a pang of regret. Not because of the sacrifices I made, but because the way I chose to live them.

Medical school was not what I expected. In Spain, our program is six years long: two years dedicated to the basic sciences, one transitional year and three clinical years. Admission is straight from high school, with applications graded solely on academic performance. Those with the highest marks get in, and I was one of those "high achievers." I was lucky; my father's job as a diplomat offered me the luxury of excellent schooling. I had no need to prove empathy, resilience or even interest in the pursuit of medicine; being the quiet diligent student in the front row was more than enough.

My first few months were filled with the promise and excitement of college life: of self-discovery; of parties and exams; of friendships and romances. I felt the world opening up around me. And for a while, it stayed open. I recall that year fondly. The lazy afternoons playing cards in the cafeteria, my first boyish crushes, the quiet trepidation of stepping into the dissection lab for the first time ... I was unaware of what I had gotten myself into, but I was enjoying it. After all, biology and philosophy had been my favorite subjects in high school. Medicine seemed like a good fit, the intersection between the messiness of human life and the sturdiness of modern science.

Second year arrived, and with it the toughest subject in our school: anatomy. Infamous for weeding out the "weaker" students, we had all heard the horror stories. I paid little heed, too busy discovering the pleasures of university life. When finals came, I studied, and failed. And then failed again. Fear made me begin to skip classes and even exams, and I was held back a year. My classmates were replaced by strangers, and my friends moved on to the far away world of the hospital. I felt alone. The openness of college was replaced by deadlines and exam dates. Stuck in the hamster's spinning wheel, I had direction but not purpose.

Instead of facing my failures, I channeled my anger and frustration into student representation. Nights fretting over the next exam turned into endless meetings, projects and emails. And for a while, it worked. I found meaning in the work that I had not in my classes. Cell membranes and tendons were too abstract, too removed from the world of medicine. Fighting for better classes and fairer exams was not. Not only did I sublimate my feelings of failure into activism, it gave me an excuse. I wasn't failing because of me, I was failing because of our subpar education. And while I was partly right, I was also very wrong.

Being held back was eventually replaced with the very real possibility of being kicked out. Student representation was no longer able to shield me from reality. *"I know it's tough to swallow, but you have to study to pass, not to learn,"* I was told. So I changed. I abandoned the meetings and emails. Medical school became a crucible, my passion smelted into pure survival instinct, all else discarded as slag. Becoming a good doctor was no longer on my mind. Textbooks and Medscape gave way to high-yield notes and exams from past years. Memorizing the nigrostriatal pathway had less to do with understanding how to help my patients with Parkinson's disease than with passing neuroanatomy. And pass I did.

I limped from subject to subject, until I arrived unceremoniously at the end of my third year. The worst was over. I was no longer on the cusp of being expelled. I had survived. And all that was left before me were the "best years" of medical school, the clinical years. Textbooks would give way to patients, diseases and the heroic world of medicine.

I remember asking a fellow upperclassman for advice during clerkships. Her words stuck with me: *"Take your closest friends, put them in a shell, and don't look back."* In my naiveté I rejected the idea. *Medicine is built upon helping others, cooperating for the greater good of our patients. How can I become a good doctor if to do so I must reject the basis of our craft?* The very notion disgusted me. I promised myself that I would avoid that cynicism; I would be different.

The clinical years started, and I was finally learning about medicine. Not cell membranes, but diseases and cures. Most of our clerkships consisted of glorified observerships, but I cherished the contact with patients. The exams were tough, yet nothing compared to what I had gone through. *"Maybe this is why they were so tough on us,"* I thought. I soon settled into a routine, sliding from exam to exam and clerkship to clerkship. I kept my friends close while my classmates drifted away. When the passing mark depends on how well your colleagues score, you start to see them as rivals. *Why should I help them? What do I gain in return?* Cooperation was replaced by suspicion. And it helped me succeed.

My sixth year and final year began with a bumpy start. While previous clerkships had always been second in importance to exams, this year it was all about rotations. "You'll be treated as interns," they told us. And they were right. It's easy to avoid bullying when doctors ignore you; not so much when you're supposed to be "one of the team." It took me several months to recover from that first clerkship, but I survived. I had become an expert at distancing myself. Surviving rotations was not that much different from surviving exams. All it took was strengthening that shell and powering through.

Christmas break came around, and with it the prospect of graduating in a few short months. I did not feel excited or hopeful. I felt empty, even fearful of what residency would hold. Something was missing. That's when the realization hit me: I had taken my friend's advice. I had cut myself off from the things that gave me meaning as a way to defend myself. Instead of flourishing, I had concentrated on surviving. And I had done a good job. But it had left me wanting. While I now knew how to treat a plethora of diseases both common and rare, I had forgotten why it was worth treating them in the first place. Delayed gratification had transformed into delayed growth. The true tragedy of our education is not its lackluster quality or even the institutional hazing we endure, but the sundering of medicine and meaning.

Marcus Aurelius wrote in *Meditations*: "People cut themselves off — through hatred, through rejection, and don't realize that they're cutting themselves off from the whole civic enterprise." Even if we distance ourselves as a way to avoid pain, we unwittingly remove ourselves from the things that give us meaning. I still struggle to separate my own cowardice from the shortcomings of our education. But with depression and even suicide among medical students on the rise, we must reflect on the type of culture we are imposing. How can we expect to care for our patients if in the process we are forced to hurt ourselves?

In a little over a year, I'll be a full-fledged intern. I will have the privilege and the burden of treating my fellow man. In the past months I have found passion and meaning in medical education, in ensuring no other student has to go through what I did. But is that enough to be a good doctor? Why does medicine hold meaning to me? What does being me mean? These are questions I have yet to find answers to. With the ice age of medical school thawing, I recall the lyrics of St. Vincent's song "Ice Age": *the answer is close to my bones, and far from my shell.*

I do not regret choosing medicine. But I do regret living these years bereft of purpose. Moving forward into residency, I am inspired by Paul Kalanithi's reflections. Paul was undergoing his final years of residency training when he was diagnosed with stage IV lung cancer. Uncompromising in his mission, Paul chose to dedicate his last months to the search of meaning. His words reminded me that there is more to a patient's disease and suffering than symptoms and drugs. That in helping my patients find meaning in their journey, I can find meaning too.

As I began medical school, a physician I now consider a mentor told me that the miracle of medical training is that it asks you to become a person who can move forward without being stopped by remarkable challenges. This is not to say that one should not reflect and process, or that this aspect of training is unique to medicine. Rather, it points out that the presence of these obstacles is not left to chance: as a trainee, you will almost certainly become attached to a patient whose course crushes you, awkwardly step into a family who has just lost a loved one, wonder if there was something you should have realized when it is your own family that experiences loss. And then there will be more to do.

The first year of clinical rotations — classically the third year of school — is the first year that many students experience this pace. The pieces in this section are among the closest to my heart. Emily Fu and Parsa Salehi grapple with blurred ethical boundaries. Chris Meltsakos, Rohini Bhatia, Ashley Ellis, and Diane Brackett recognize in themselves a new lens to view death and dying. Elaine Hsiang and Malone Hill describe exhaustion, the literal and figurative reminder to check our own pulses. Each of their stories is humbling in its own right, marking a transition into a new set of hardships and healing.

Clinical Rotations

Gladiolus by Paulette Pham
Class of 2018 at University of Texas Medical Branch

Clinical Rotations:
Ethics

Medicine in Translation

November 18, 2016

Emily Fu
Warren Alpert Medical School of Brown University
Class of 2015

D URING MY LAST VISIT home, my mother waited less than an hour before showing me her medical records. She offered them up the way I'd once presented my middle-school report cards, steering the papers across our kitchen table between bowls of peppercorn chicken and eggplant until they slid to a stop in front of me. Looking at them made my head spin, as they were written almost entirely in Chinese.

They were from the National Taiwan University Hospital, where my mother had been undergoing treatment for breast cancer for the last two years. She discovered the lump the summer of 2015. A month later, she moved into my grandparent's house in Taichung, leaving my father and me to worry from our house in New Jersey. When asked about her reasons for seeking care outside of the United States, my mother always cites Taiwan's single-payer, nationalized health care system, widely acknowledged to be one of the best in East Asia. She notes that her co-pay for office visits never exceeded twenty dollars, and her paperwork was minimal despite numerous referrals. She's too pragmatic to discuss anything beyond the logistical benefits of her choice.

Still, I think she was also motivated by her desire for caregivers who wouldn't judge her for her accent and her beliefs. She wanted nurses who would recommend congee for her lost appetite instead of cereal and doctors who wouldn't balk at her cupping marks. After having faced everything from casual bigotry to English grammar during her thirty years in the United States, she wanted care from those who were fluent in her language and culture. When her cancer went into remission and she decided to return to America for her follow-up care, I assumed the hard part was over. I was wrong. Transferring care from an international hospital to an American provider was — and still is — a painful process, driven by a patient's skill at jumping through bureaucratic hoops. This became clear when my mother's medical records stumped doctor after doctor.

Too many clinics and hospitals did not have a Mandarin interpreter on staff. One oncologist simply returned her photocopies; another asked if she could send in her biopsy. Several doctors wanted her to simply repeat, at great personal cost, every lab test and scan she'd undergone in Taiwan.

Finally, an uneasy solution was found: my mother was asked to translate the Chinese portion of her records into English and submit those instead. My mother, who was already anxious about finding an American doctor, who asked me what neuroscience meant when she saw the word on my diploma, had to translate her own medical records — which meant so did I.

It's true that my mother's diagnosis has improved my Mandarin by leaps and bounds.

"Isn't English the universal language of medicine?" she demanded. "And isn't English your first language?" Implicit in these questions was another: "Have I raised you in America and tolerated your strange bicultural ways for nothing?"

She couldn't believe that medical English was so different from American English. Months into medical school, I was still struggling to use, understand, and even spell terms like *orthopnea* and *dysmenorrhea*. Every time I described a patient as sweaty instead of "diaphoretic" in front of my doctor-mentor, I felt the flushed embarrassment of a child who's forgotten their talent show routine.

At first, my fresh-off-the-boat confusion about medical terms seemed fine, even funny. The impenetrable nature of medical English was a matter of no urgency. It wasn't until my mother's return to America that I realized how frighteningly tall the language barrier looks when one's well-being is at stake. It wasn't until I started interviewing patients that I realized how ubiquitous this language barrier is.

During my time in the emergency room and primary care settings, I've met a number of non-English speaking patients. I meet younger relatives who translate for older patients and it reminds me of my family. When the interview balances, precariously, on the use of simple English, I leave feeling queasy. Once, I joked to a patient that the only Spanish word I knew was the word for flooding: *inundación*! I'd learned it at a medical school language class that felt like a small solution to a systemic problem. The joke was meant to be self-deprecating, and to put her at ease. But when she frowned and told me I should practice my Spanish, I felt a wave of shame. Of course, self-deprecating humor was scary when her caregiver's ignorance had consequences for her care.

Luckily, most of the doctors I've met have been kind, if ineffectual, about language barriers. They're game to use devices to improve their patient encounters: inarticulate translating computers and interpreter lines reminiscent of a grade school telephone game. But, other physicians are impatient, bordering on cruel. A teacher of mine once complained that having to use an interpreter delayed her entire day's schedule and slowed down her thought processes.

While she was the most dismissive example, I've seen other physicians — kind, well-meaning ones — make other assumptions of non-English speaking patients. They speak to them as if they're deaf, dumb or indigent. The assumption, I suppose, is that patients who struggle to speak English will also struggle to understand medical concepts.

So, as a solidly mediocre medical student who has tried scaling the language barrier from both sides, I'll expand on that assumption: *everyone* struggles to understand medical concepts. Regardless of cultural context, medicine is itself a foreign language that no one is born into. I've met patients, brilliant and resilient, who cannot understand words like *benign* and *screening*. No matter how fluent they are in any language, medicine plays by its own linguistic rules — not to mention that stress and bad news short-circuit communication skills in the best of us.

Not everyone needs an interpreter, but almost everyone needs interpretation — of test results, medical terminology and esoteric hospital policies. The further I progress in medical school, the further I stray from being able to fulfill this need. I'm much better now at understanding the things that doctors tell each other outside of the patient rooms. But each day I'm in school, listening to lectures and highlighting packets, is time spent in an immersion program. I default to acronyms and complex scientific explanations. The longer I spend listening to lectures, the worse I become at speaking plain English or even bad Chinese.

Even before my mother became furious and disappointed at my inability to help her translate, I knew my education would distance me from my family. We were already separated by cultural, generational and linguistic gaps: I speak to my parents in stilted Mandarin and they reply in limited English. Medicine was yet another language barrier — how would I even begin to explain to them the smell of formaldehyde or the contents of my classes?

However, I never expected my education to distance me from my patients as well. But, every time I'm in a care setting, I have to remind myself to put aside the jargon and the casual tone with which lecturers refer to debilitating illnesses. Even as someone who has had to balance two cultures my entire life, the transition between patient interaction and physician discourse is the most challenging code switch I've ever experienced. Yet my education has done little to teach this transition, let alone acknowledge it.

People — and especially doctors — will always struggle to communicate across cultural differences. What we fail to realize is that medicine, in and of itself, is a foreign country with its own tongue and customs. If its visitors, our patients, are struck speechless and lost, it should not reflect poorly on them.

And, if those visitors are non-English speaking, they have to contend with infinitely more strife: a sociopolitical climate that is hostile to them, a country in which power is inversely correlated with accent. If those visitors are like my mother — or, as is often the case, less financially and socially privileged — by the time they arrive to our clinics and hospitals, they are already exhausted because they've had to carry the burden of communication for as long as they've been in this country. They've had to navigate American signage and systems because health care is not the only field with a shortage of linguistic inclusion. This navigation is draining, and being ill is draining. Yet we still expect these patients to communicate their medical needs by wrestling with an uncooperative language. We are the ones who have had the privilege to spend our days studying jargon. We should be responsible for bridging the communication gap intrinsic to our profession.

Hierarchy in Medicine: Compromising Values for Honors

July 4, 2016

Parsa Pamenari Salehi
Drexel University College of Medicine
Class of 2017

> "I'm not going to stop the wheel. I'm going to break the wheel."
> —Daenerys Targaryen, *Game of Thrones*

—

A UBIQUITOUS HIERARCHY pervades all levels of medicine. Medical students are anchored firmly at the bottom of medicine's social ladder, rendering them functionally powerless. Although students theoretically have a "voice," their precarious position low down makes them apprehensive to use it. Students' grades, evaluations, recommendations, etc. — which have real, tangible impacts, not only on students' academics, but also their future careers and lives — are contingent on appeasing those higher up on the so-called social ladder.

This hierarchical construct in medicine is the single most important ethical challenge facing medical education today. The reason being, it can be considered the origin of all other ethical dilemmas in today's medical education milieu. For instance, a student may feel pressured by an attending physician into unethical conduct, such as doing a rectal exam on a patient under general anesthesia without informed consent. Many residents and students fail to report working over the legal 80-hour limit, as they fear the consequences of angering a program director. Any medical student can recall an incident when they witnessed an archaic procedure, medication or test being ordered, and decided to remain silent, rather than challenge an obstinate attending. Perhaps more disturbing are the sundry anecdotes of students suppressing their right to defend their own personal beliefs (religious, racial, political, etc.) in the face of prejudice, when the person holding the bigoted views is an individual high on the medical hierarchy.

The aforementioned list of ethical dilemmas is far from exclusive — a multitude of ethical quandaries arise from medicine's current ranking framework. At its crux, the issue is that medical students must continuously accept coercion and compromise their personal values, ethics and mental health, or risk challenging the hierarchy that dictates their future careers and lives.

It is hard to imagine medical students alone shifting a pervasive paradigm in medical culture. However, one day we will hopefully all be residents and physicians, and it is at this point where we must be prepared to take a stand and alter the current construct. In order to accomplish this, we, as a community, need to first elucidate the concept of medical hierarchy and its impact on medical student learning and functioning. Second, we must acknowledge and discuss the inherent ethical dilemmas this type of construct creates. Finally, we ought to continually search for diverse solutions to this challenge. An effective way to accomplish this goal is to look to proven models from other professions as examples. For instance, at Facebook Mark Zuckerberg regularly holds town hall-esque meetings during which any employee, regardless of their position, has the opportunity to ask him questions, voice suggestions and showcase innovative ideas. Although a hierarchy still exists at Facebook, simple practices such as this foster creativity by bolstering camaraderie and confidence in the workforce. Such systems do not eliminate the ladder. Rather, they simply decrease the distance between the rungs.

In fairness, I am far from an expert in the aforementioned predicament. However, I do firmly believe that if we value this topic within the medical student community, we will be ready to make positive changes in the future. Ironically, for the benefit of medicine's disciples, we will have to use the very "power" granted to us by the medical hierarchy — as residents and attendings — to positively transform the system.

In summation, unlike Khaleesi, we do not need to "break the wheel." Medicine needs structure. There must always be capable mentors sitting at the top of the pyramid, guiding the students below them. Nevertheless, with our collective brainpower, I am convinced that we can come up with a more successful system. A system that is not only better for all healthcare professionals — from the lowly medical student to the attending — but will also improve patient care, foster a collaborative environment and promote innovation. We do not have to, nor should we, break the wheel. We just need to figure out how to roll it in the right direction.

Clinical Rotations: Death and Dying

From Birth to Death:
A Recollection of the Third Year

July 22, 2013

Chris Meltsakos
New York Medical College
Class of 2014

U PON ENTERING MEDICAL SCHOOL, we all knew that we would have to deal with some difficult diagnoses, emotional situations and even death. In fact, even the earliest portions of our training were centered around a cold, lifeless cadaver that we cut into to learn the intricate anatomy and beauty of the human body.

To a first-year medical student, gross anatomy symbolizes the profound meaning of what it is to embark on the long journey of becoming a physician. When one first picks up that scalpel and sets his or her eyes upon the pale skin, with the pungent odor of formaldehyde filling the room, an image is engraved in his or her mind that sticks with them for a lifetime. With the amount of time that is spent working with one's cadaver, one is bound to wonder about that shell of a person that lays before his or her eyes. One recognizes that this body holds the story of a life once lived, bearing life's scars and the wrinkles of time and gravity. It is most medical students' first real encounter with death beyond attending a funeral.

Yet, as profound as that first experience with the cadaver may be, there are no books, no lectures, no seminars and no words that prepare us to embark on our third-year clerkships. It is a year that starts out with such excitement and flies by quicker than one can believe, yet during this time period, one sees himself or herself grow and mature more quickly than he or she ever believed possible. This transition from a student to a caregiver does not happen overnight. Sometimes it does not even happen until the third-year clerkships are almost over.

Yet for each of us, this transition occurs. Somehow, all of the initial anxiety and nervousness that floods one's mind on the first day of clinical responsibility seems to fade and is replaced by a feeling of "Hey, I can do this! And I'm not half bad at it either!"

But what is it *exactly* that that fosters this transition? Well, in my opinion, it is a combination of a number of factors. There is great responsibility put on each of us as we progress through the third year. You are now both a student and clinician trying to balance enough patient care with reading and doing well on exams. It is a year in which one feels overwhelmed, overworked and utterly stressed out on many occasions, but somehow each period of stress or anxiety is offset by wonderful memories and experiences gained from both our peers, who work closely beside us, as well as the individuals who probably teach us the most: our patients.

My third year began with my pediatric rotation. I did not know what to expect, except that I knew that I was not interested in pursuing a career in pediatrics. As the rotation went by, I saw everything from babies seizing from herpes encephalitis to children bedridden and bald battling acute lymphoblastic leukemia.

There was one patient who changed my life forever: an adolescent girl, who was in a head-on motor vehicle accident, sustaining bilateral femur fractures, a radius fracture and pelvic fracture. But more importantly, this was a little girl who watched her mother take her last breath at the scene of the accident. From the first day she was on the floors as my patient, we had an unspoken bond. I stayed many late evenings in the hospital helping to ease her worries and concerns and simply talking to her about anything that comforted her. It was not only a moving experience, but it was an emotionally trying experience. On the one hand, it was amazing to see this little girl improve and get better. On the other hand, my heart was filled with such pain and sorrow watching her cry and tell me about the relationship she had with her mother.

That is when I learned one of my biggest life lessons: The hardest question that a patient will ever ask a physician does not involve the mechanism of action of a drug or the pathophysiology of their condition. Rather, it is the question, "Why?" Whether they say, "Why me?" or "Why my mom, dad, sister, brother?" This question is the hardest question to both answer and to wrap one's mind around.

As my third year progressed, I saw teary-eyed families surrounding comatose patients as we "pulled the plug." On an overnight call, I had my hands deep inside of an abdomen assisting with a ruptured aneurysm, when the patient expired in front of us. I helped to complete bowel resections and to remove tumors. I delivered babies and saw babies die before they could even experience an hour on this earth. I watched a 48-year-old mother of three with pulmonary adenocarcinoma take her last breath in front of her family as they cried verses from the Bible. For every third-year medical student, the list of life-changing experiences is expansive, with each person finding strength and meaning in different connections that we make or lose.

My third year ended with my internal medicine clerkship, one of the most difficult clinical experiences of my career. We had just admitted a 74-year-old male who was complaining of dyspnea and generalized weakness. Several months earlier he had been told that the B-cell lymphoma that he had been receiving treatment for was in remission. However, six weeks later, when he came in for a routine PET scan follow-up, his oncologist found that the tumor had recurred and was now the size of a small grapefruit. We hospitalized him for a congestive heart failure exacerbation, giving him the usual Lasix and contacting his oncologist to visit him in the hospital to administer his chemotherapy regimen.

The patient's oncologist asked me to assist him in relaying to the patient that, despite his chemotherapy, he was most likely going to pass away within the next year. Entering the room with one of the oncology nurses, we explained to the patient that his recurrent cancer was not only aggressive, but also that his comorbidities put him at further risk. If his cancer didn't kill him within that year, one of the other issues could.

First there were a few seconds of painful silence. Then, he asked the nurse some questions about his chemotherapy before asking if he could just talk to me. My heart, having already sank from the initial difficulty of watching a person respond to being told that he only had a year left to live, felt like it was in my feet. I was nervous, scared, sad and wondered, "Why would he want to talk to me?"

I sat down on the patient's bed beside him, as I always did each morning when I rounded on him and said, "I know that this must be hard. I can't imagine what you're going through, but I just want you to know that we are all here to help you every step of the way." That's when he said to me, "You know what, doc? I'm not even scared ... How am I not scared? I'm about to face God and I'm not scared? Do you think I should ask for forgiveness now? How do I prepare myself to die?"

I paused for a moment to collect my thoughts. Slowly, trying to spit my thoughts out, I began telling him that it is normal for people to be numb at first. I tried to address his questions about facing God by telling him that we each have our own connection with the creator we choose to believe in and that he needs to follow what is in his heart to bring him the most comfort. He then proceeded to say, "Well, doc, what do you think it's like, you know ... right before you die?"

Stunned, I tried to collect my thoughts and after a few moments responded with, "Well, I suppose I don't know. I mean I honestly don't know. I don't think anyone could tell you that, which is why it's such a scary thought."

We continued this conversation for over an hour, talking about both life and mortality and how he could tell his family members. Over the next four days I spent around an hour with him each day, talking about anything that was on his mind. Some days were tearful and difficult, and other days were a bit more lighthearted. One day we talked about whatever food was being featured on the Food Network.

When the patient was ready for discharge, he gave me a big hug and said, "Doc, you're my savior. Thanks for talking to me for all of these hours. I'd have gone crazy otherwise. I know ... I won't probably see you again before I die, but ... I'll watch from up there." He pointed up towards the ceiling. "Good luck with your career. You'll be great."

There is nothing to prepare a medical student for the experiences of third year. One will laugh, cry, feel overwhelmed and stressed. One will feel pride, awe and happiness. Maybe more importantly, though, one will witness the full circle of life from birth to death. The beauty of life and the human body make it a true privilege to be a part of this field. These experiences not only change us as people, but they help shape us into the physicians that we will become. In the words of Sir William Osler: "He who studies medicine without books sails an uncharted sea, but he who studies medicine without patients does not go to sea at all."

Oceans Away

February 24, 2017

Rohini Bhatia
University of Rochester School of Medicine and Dentistry
Class of 2019

I AWOKE TO A PHONE ringing frantically — must have been a Whatsapp call. My father yelling from downstairs: "He passed." And my mother, opening my bedroom door before my eyes had fully opened, who stood there with her cellphone out, lips quivering, and eyes searching: "He's gone." My grandfather had passed away.

The month of December had been tough. Earlier this month, my *nana* (Hindi for grandfather) was hospitalized due to swelling in his foot. As hospitals tend to do, they worked him up for everything under the sun. His hoarse voice that he'd had for about a year was now a symptom under scrutiny. Bronchoscopy, laryngoscopy, Foley catheters. Every night (U.S. time) through Whatsapp phone calls with my uncle, my mom would relay the procedure of the day. And every morning, we would listen for the results.

My family knows I am in medical school — but their vision of what I know is very different from what I actually understand. Despite this, pictures of prescription pads and notes, lab values, and X-ray interpretations were texted to us. My father and I exchanged ideas and hypotheses about what could be going on. Just a week earlier, on the road to a family Christmas party, we received a Whatsapp call. "TB?" my mom had exclaimed with exasperation. *What on earth.* How peripheral edema resulted in a hospital diagnosis of tuberculosis was beyond my comprehension. *"Was it really TB...?"* was the first thought that went through my head. "Text us the results of the acid-fast or an image of the chest x-ray." It's got to be a misdiagnosis, right? But, who am I to question an experienced physician's interpretation?

My mind whirled with images of the SketchyMicro TB sketch. RIPE therapy of course. He should probably get an ophthalmology appointment before starting that darn ethambutol. But when we got the list of medications he was given, I recognized none of the names. Oh, India. Urimax instead of Flomax. R-Cin for rifampin. Luckily, Google works for translation of pharma names. As the days went on, I kept trying to develop a story in my mind for what was happening oceans away at a hospital in Bombay. They had put my 86-year-old grandfather into an isolation room and told him that "this was so you don't get infected further." My mother said he was no longer reading the newspaper, making her extremely nervous. Normally, my *nana* would wake up every morning, drink chai and sift through the three different newspapers that arrived at their Bombay home: Midday, Hindustan Times, India Today. His isolation room and his lack of awareness at what was going on outside of that worried me. My grandmother, my *nani*, was also worried. He was beginning to lose strength. He couldn't pass urine and hadn't had a bowel movement in days. "But it's just TB," I said to my mom. "It's completely treatable." She was booking a flight to India to help with his care. "No need to rush," I said.

That morning, I helped my mom pack for India. The first thing she pulled out of her closet was a white *salwar kameez*. For the cremation. I tried to be reassuring, "He died of a heart attack. No one could have predicted it." I think that was more for myself. I couldn't help but dwell on what I had missed.

It seems that the art of determining severity of a medical condition was beyond my capability. Had I missed something? TB is treatable, right? How severe was it? His heart was doing fine ... He had a coronary artery bypass graft 30 years ago, but no real problems since then. Were they related? Or was it the fact that he was in the hospital? So often we hear from doctors that the hospital is the most dangerous place to be. Of course, some of this notion is related to the fact that admission to the hospital is often due to symptoms that are seen or felt, an indication that there was a tipping point in the disease process towards something gone awry. Yet, I couldn't shake the feeling that there must have been something else. If *nana* hadn't gone to the hospital, and if he had still had this heart attack, he would have lived his last couple of weeks at home, not in an isolation room. The logic in me understands that there is obviously a need to impede spread of infection by an individual affected by TB — but what if it wasn't TB? What if it was a solitary nodule with TB that had just been walled off? What if all of this was unnecessary?

My regret is not seeing the seriousness of it earlier. I don't know if I could have, but if I missed signs, that is what terrifies me. How do we navigate the waters when a loved one becomes ill? More so, how do we navigate the waters when there are oceans of distance apart and you are interpreting and speculating through secondary data? Throughout his hospitalization, I looked at my *nana's* illness as a solvable puzzle (they'll probably put *nani* on isoniazid, you know, for precautionary reasons), but I never actually spoke to him. I got caught up in the medicine, and even that wasn't helpful.

The Hidden Challenge of Medicine: The Art of Sharing Bad News

March 1, 2017

Ashley Ellis
University of Kansas School of Medicine
Class of 2018

M Y ALARM WENT OFF at 4 a.m. in the morning. I begrudgingly pulled myself out of bed, threw on some scrubs, and headed to the hospital. Not a car was on the road. It was the third week of my OB/GYN rotation, and I was on the infamous gynecologic oncology service. Rounding began at 6 a.m. sharp, and I needed to first check in on my post-op patient from the day before. She had undergone an extensive bowel resection for metastatic ovarian cancer and I knew her prognosis was grim.

Once I arrived to the hospital, I glanced over her chart, per usual, and headed to her room. Despite my waking her, the elderly woman kindly allowed me to examine her. She was frail, with matted hair and a hollowed-out face — far from the picture of health. She grimaced in pain as I softly felt her abdomen. Despite her obvious discomfort, she continued to smile weakly throughout my exam.

As I prepared to exit the room, she questioned me about the results of her surgery. It caught me off guard that no one had talked to her about this already. As a medical student, I'm far from a cancer expert, but even with my limited knowledge, I knew this patient was dying. It dawned on me that this woman did not have good insight into her condition. Consequently, I began to fumble my words and my palms began to perspire. Feeling rather uncomfortable, I replied that I wasn't sure of the details, and that the physician would be by within a couple hours to tell her more.

As promised, an hour later our team entered my patient's room. The woman's daughter and pastor were now present. The surgeon began to explain how we had removed a vast amount of tumor from the patient's abdomen and commented that this would likely relieve a lot of the belly pain she had been experiencing. The woman seemed very pleased to hear this news and exclaimed, "So I'm cancer-free?"

"I'm afraid not," the surgeon slowly replied. She went on to describe how the cancer had spread throughout the patient's abdominal cavity, like little freckles coating her intestines. It had latched onto her diaphragm and invaded through her abdominal wall. The surgeon presented the facts in a forthright manner, yet her face remained soft as she knelt down to the patient's level. The surgeon looked directly into the patient's eyes and grasped her hands as she explained what all of this meant. Despite their best efforts, this was not curable cancer. The patient's face sank in disappointment. I could sense her daughter's confusion, trying to process this news. "So how long does my mom have?"

Providing a precise timeframe for such a complex disease is daunting and a task that medical professionals oftentimes dread. One can never predict exactly how a cancer case will unfold. As a third-year medical student, this was my first time witnessing such an encounter and I wondered how the surgeon would reply. As the daughter asked this question, I felt my heart flutter. My body froze and I held my breath in anxious expectation. Was it best to answer vaguely or was it better to be as honest as possible, given the surgeon's knowledge of diseases of this nature?

After a brief pause, the surgeon replied in a matter-of-fact tone, "I would estimate about six months." The sentence hit me like a rock; I can't even imagine how it must have felt for the patient or her daughter. This was obviously not the answer that they were expecting. The daughter immediately became hysterical and ran from the room. The patient's face conveyed helplessness and fear. Teardrops began to slide down her cheeks. Her mouth hung open, yet no words came out. The following 45 minutes were a blur. We sat with the patient, talked through her emotions, handed her tissues and discussed what her goals were for her remaining time. I stood back, watching the resident squeeze the patient's hand. The physician offered words of encouragement, but I honestly do not remember much of what was said. My mind reeled with pain and sadness for this woman.

As I progress through my career, I hope to never forget this encounter or the people in that room. Although some physicians may have shared the painful news differently, I admired the strength of the physician as she led that conversation. I believe that family needed a direct answer, and so the surgeon gave it to them. I truly think these conversations embody the most challenging part of medicine, so I felt honored to bear witness that muggy July morning. Although I'll never know how the rest of her life played out, that patient will forever be in my thoughts. People who aren't in medicine often tell me they can't imagine memorizing all the information I must know to be a physician. I now tell them that learning the science of medicine is far from the hardest part.

After the Autopsy

May 13, 2016

Diane Brackett
University of Central Florida College of Medicine
Class of 2016

I WENT THROUGH MEDICAL SCHOOL without experiencing the death of a patient I had personally cared for. In contrast to what may be seen on the trauma service, my surgery clerkship was full of routine procedures: appendectomies, cholecystectomies, port placements, excisions of pilonidal cysts and miscellaneous "ditzels," as pathologists may refer to them as. Sure, I have had patients who were quite sick and did not have much time left to live. For example, I once performed a neurologic exam on a comatose teenager in the ICU, whose arteriovenous malformation had bled wildly out of control despite prior neurosurgery. But with the constant shuffling of rotations that medical students must endure, I was always in and out of patients' lives before they had a chance to leave mine.

On my OB/GYN rotation, I bonded with a patient who had metastatic ovarian cancer. During her surgery, I peered helplessly over her wide-open abdominal cavity as the surgeon grabbed fistfuls of mucinous tumor and tossed them onto the Mayo tray. My heart sank when she asked me the following morning how her surgery had gone. She was a friendly woman who boldly faced her impending death. She told me that if there was a heaven, she would be looking after me. Knowing that my rotation was about to end and I would not be able to follow up with her, she asked if she could send a letter to me, once she was out of the hospital, to let me know how she was doing. Though it crossed my mind that this could be a breach of professionalism and the doctor-patient relationship, I graciously gave her my address. I never received a note.

Thus, I cannot speak from experience how I would react to a patient dying — someone who was speaking to me one day, dead the next. But I've learned how I may react to a patient who is already dead. I'm not talking about my anatomy lab cadaver, who was my first patient. I'm talking about a patient who was alive mere hours ago. As a newly matriculated medical student, on the eve of our first anatomy lab, I had been fearful that my cadaver would be too lifelike, tricking my mind into thinking his eyes might flash open at any second. Instead, my impression was that he was *so dead*. I poked, prodded and dissected in relative peace, without any self-imposed hallucinations that he might suddenly wake up. Fast forward to fourth year. The hospital autopsies I have participated in all involved recently deceased, unembalmed patients — a far cry from the cadavers, who had been embalmed not once but twice, and for all I knew, were stowed in a cooler for up to a year before being unveiled to medical students. Although the cadavers played an important role in introducing me to death, autopsy pathology has pushed me further, with new challenges and greater demands for adaptation.

My first autopsy was hard enough, but I accepted it. The patient was a middle-aged woman who had a number of serious, chronic medical conditions that were thoroughly documented. Her death was not very surprising. I have also assisted in autopsies of fetuses, stillborns and neonates who died shortly after birth. Though sad, I accepted that these babies had terminal conditions and did not have to suffer for long. They were like little hatchling sea turtles, snatched away by hungry birds before they had a chance to make it to the ocean waters. I saw it as a part of nature.

It was the autopsy of a young man, barely out of his teens, that had me doubled-over, grappling with life, death and the unfairness of it all.

He was short-statured, with thin, lanky limbs, dark hair and a farmer's tan. As far as I know, his illness sprung on him unexpectedly, and he didn't know he was going to die. I puzzled over his face. His expression, for the most part, was peaceful. When the endotracheal tube was eventually removed, his lips remained slightly parted, as if he had dozed off, causing his jaw to relax. His eyes, however, were shut with a subtle squint, as if he was wincing. I imagined he was still present in his body, trapped and aware ... something akin to locked-in syndrome, but worse, with no ability to move his eyes. His thoughts were racing as he pleaded to know what had happened to him. And I was the only one in the room who could hear him. His struggle became more real when I went to pull his arm away from his body, so that the resident could get a better view while taking photos. Though I knew his stiff arm was simply a postmortem effect, I still imagined he was resisting us.

Goosebumps covered his arms and legs. My first instinct was that he was cold and uncomfortable, the same way I would be if I were laying naked on a metal table in an air-conditioned room. My rational side snapped back at the illusion, remembering from a recent forensics lecture that the arrector pili muscles are also subject to rigor mortis.

With the completion of the external examination, the diener stepped forward with a scalpel and swiftly sliced open his body. Of course, death is final, permanent, absolute. Black and white. Yet this man's death was becoming even more profound with each organ that was removed. After the initial Y-shaped incision, I rationalized that a person could still survive that insult. Many sutures would be required to repair the lacerations, and a good deal of narcotics would be needed. And the heart ... well, there are people walking around with heart transplants. Technically they were alive for a brief time without a heart. But then came the electric saw, the chisel and the bone-cracking that signaled the removal of the skull. *Snip, snip, snip ... snip ... SNIP!* The brain was pulled clear. In my mind, there was no return. Maybe somewhere within the dead and dying neurons, chemical fragments of this kid's memories still existed and were still residing in his body. But now even that was taken away and stored in a bucket of formalin. I contemplated the essence of personhood. Was he still a person? Or had we turned him into the shell of a person? I thought forward to his memorial service. His family was planning on having an open casket. Who would we be fooling? Are people actually fooled, or do they just hope and beg to be fooled? My medical training has permanently changed the way I see things. I will never be able to go to an open casket funeral without picturing the hidden "pillow line" suture, running from behind one ear to the other. I've learned that this technique allows even a bald man to have his brain eviscerated and still keep up appearances.

At home, I searched online for our patient's obituary. I wanted to know who he was, to acknowledge that he was a person and not just another autopsy case, as I felt the rest of the team had treated him. I convinced myself it was okay to look him up online, because it was clearly public information, out there on the world wide web for everyone and anyone to read. But there was nothing. It was too early, I figured. In the meantime, I resorted to Facebook.

I found him. At least, I was pretty sure it was him. There were only a couple of profile photos. One had dark lighting. The other was grainy, and he was wearing large sunglasses, obstructing a complete view of his face. Everything was consistent with what I knew about him, yet I wasn't fully convinced. I was not willing to commit, as there was a chance I was mistaken. Due to the security settings of the account, there was not much else to see. I scrolled down, searching. Then I saw it: a video. Impulsively, I clicked on it. I was brought to a summertime scene in a backyard. A teenager with shaggy dark hair stepped into view. He was wearing a bathing suit and sandals. I instantly recognized him — his lanky arms, his thin, bare shoulders and his slightly stooped posture. The endearingly awkward kid I construed in my mind had now come to life. It was confirmed. Tears sprung in my eyes as I watched him proceed with the ice bucket challenge. After being doused with ice water, he laughed, grabbed a towel, and walked away. The video ended. I was glad I had left the speakers muted, so his voice and laughter did not add to the pain that was tightening its grip around my throat. As I closed my laptop and sank down onto my bed, a few more tears trickled down my face. I turned off the light and allowed myself to succumb to my fatigue.

His obituary was eventually published, and in it I found the closure I sought. Now I knew who he had been. He was a good kid. He had just finished his first year of college, and aspired to be a high school teacher. His family missed him very much. Some might argue that all this "extra information" makes these situations more difficult and painful than they need to be, but I've never been the type to shy away from wanting to know as much as possible. It was more important to me that I acknowledged the patient. I knew I would never meet his family, that they would never know I had been there or cared that much. Yet, I believed they would be appreciative if they knew. Sometimes, the thought alone is enough.

After this experience, I went on to complete an entire elective in autopsy pathology, but not once did I attempt to look up another obituary. Perhaps the first one was the only one I needed. It was the sense of preserving my own humanity that freed me to fully pursue the objective, investigational work that is required in the field of pathology, my chosen specialty. Perhaps in the future, as a resident and later as a practicing pathologist, there will be times where I'll find myself searching for a patient's obituary. Maybe that's what I'll need to do from time to time, in order to process, accept and move on from a particularly challenging or emotional case. I will certainly not be the only doctor to do so. Even in specialties where patient contact is abundant, physicians may still realize that they often only catch a glimpse of who their patients really are. In her *New York Times* article, Dr. Allison Bond eloquently reflects: "In the stream of lives and deaths we witness in the hospital, it's easy to forget that we usually are privy to only a brief snapshot of our patients' lives ... [W]hen patients do pass away, their obituaries are a gentle reminder that behind the illness lies a story and a unique human being. That's something that is easy to forget, but vital to remember."

Clinical Rotations:
Burnout

palpate

July 19, 2017

Elaine Hsiang
University of California, San Francisco School of Medicine
Class of 2020

you almost died today.
you almost brought yourself back to life

knowing another heart, too,
makes warmth between your breasts.

given the choice to react or fibrose,
become a healer.

on the next plum dusk, touch the side
of your neck. it will grow and fill the room

with books that break open themselves
cradles under dim light. and when

some part of your chest
learns too many ways to describe pain

let it wander into a flower shop
tell it to find the sweetest bouquet.

—

Author's note: "palpate" was written after a session of The Healer's Art on grief. It is a poem reminiscent of a few things, including Orlando, and of taking on the role of patient while in medical training.

Cynical Yet?
A Med Student One Year Later

April 30, 2015

Malone V. Hill III
University of Texas Medical Branch at Galveston
Class of 2017

I USED TO WORK as an anesthesia tech at a hospital in Austin, TX. I was surprised the first time a doctor asked me, his incredulous tone dripping with disbelief, "*Why* would you want to want to go to medical school?" It wasn't the last time that happened, it wasn't exactly making me excited to go to school, and it wasn't a flattering reflection of the doctors who said it, but physician cynicism about the future of health care wasn't something new to me, either. People fear change, but I think people's perceptions about impending change are shaped just as much by their perceptions of themselves, especially the interacting dynamics between themselves and their evolving environment.

Personally, I know that I have changed significantly since my time working at that particular medical center. I remember my first clinic site visit of medical school, when a mother brought her 10-year-old son to the pediatrician. I listened patiently and sympathetically as the mother recounted symptoms of anxiety, depression and anger in her son, along with subtle inferences about this family's meager finances and broken family dynamics. The pediatrician told me in confidence, after I had taken the history alone, that these symptoms were exaggerated: The mother was yearning for a quick fix, or even worse, desiring benzodiazepine medication for herself. And I believed the pediatrician.

Shortly thereafter, I wrote an essay about the patient-doctor fiduciary relationship and the need to inflict short-term harm by calling child protective services in cases where the long-term household instability could be improved by requesting intervention from state agencies. I had been influenced by the pediatrician's "diagnosis" of the family; I still believe that it is difficult, but necessary, to remove pediatric patients from deleterious environments if the situation warrants such intervention, but I didn't think of the child's family as the source of the problem until the physician suggested it. I also wrote about the need to be honest with one's patient, and how the "nature of [medical] information, as personal as it is, and the unique intellectual power [and training that] doctors possess, make honesty a prerequisite in doctor-patient relationships." I still hold this belief in utmost regard.

However, many doctors' skepticism about the honesty of their patients, and the fact that face-to-face acknowledgement contrasts so starkly with back-room cynicism, casts a shadow of doubt on my faith in honesty as it pertains to the medical profession. I've never witnessed dishonesty, and I don't label the aforementioned example as such, but not every patient encounter is a case of forthright conversation, either. While I reread my idealistic beliefs from a year earlier, I realized that I'm adopting some of the cynicism that I've witnessed in others since matriculating into medical school, but I know that it's largely a byproduct of a new environment.

I recently assisted in the care of a diabetic kidney-transplant patient at a volunteer clinic near my school. Because he was a longstanding patient of the clinic, we could review his chart and progress for the past seven years, but the long list of updated medications for various health problems made it difficult to keep track of his case. The fact that he didn't speak English exacerbated his case, but I did my best to act as translator for our small team of volunteers. The patient's voice was soft and calm, as if he was resigned to the myriad problems afflicting his body. His chart chronicled "noncompliant insulin therapy" several years earlier, which the fourth-year medical student read without surprise. In retrospect, I can't imagine the trepidation this man must have felt, waiting four hours for a group of gringo doctors-in-training to sort through his paperwork, perform exams, deliver news through a mediocre translator, and not just that day but for the past seven years, relying on such resources to manage what he knew to be a complicated, if not unpromising, case.

Little does he know of his caretaker's skepticism about the adherence to his therapy, a product of an ingrained physician belief that many patients either do not care or simply refuse to listen to their doctor. *In broken Spanish?!* Who knows if I even enunciated his instructions clearly; no wonder I recognized the resignation in his voice. Yet we come to believe what we experience, that diabetic patients are "noncompliant" (it says so in his chart), probably because they're lazy or don't care, right? This patient experience came after I wrote about patient autonomy in yet another essay. I talked about how the best possible patient care, however subjective that concept may be, can be defined by the patient's conferred knowledge and his informed decision, a responsibility that entails the doctor foremost. What I didn't address are the gaps in communication, half-truths, and patient-doctor skepticism that have permeated through the consciousness of both parties, intentional or not, and how self-perpetuating these beliefs can be. I recently saw an ethics article wherein Dr. Hébert wrote, "Lack of candour or outright deception, even when well intentioned, can undermine the public's confidence in the medical profession." No, I do not believe that doctors are deceitful, but our attitudes have shaped our perceptions, and Hébert's words ring like a warning rather than an observation.

I'm sure many students have taken opportunities like this to rant or focus on the pessimistic drift they've begun upon since entering medical school, yet I don't see the need to conform to such an attitude, whether its after-effects can be seen in the older physician population or not. If I hold autonomy among the utmost of covenants in patient interactions, then I will value that same level of importance for myself, avoiding popular belief, and shaping my own perceptions. Change is inevitable, but it isn't required after nine months of medical school. One year ago I said that my school's white coat ceremony made me feel no different in terms of my professional or personal development, and honestly, I still feel that same way. I change slowly, but I hope I'll acknowledge that change, recognize it, and avoid its many pitfalls. French author Francois de La Rochefoucauld once said, "The only thing constant in life is change." That is only one of the many reasons why older doctors questioning my desire to join their profession seemed so outrageous, because like everything, the essence of being a physician is changing. Unfortunately I dread some of the unavoidable changes that I will undergo by their age, a product of my perceptions molded by the dynamics of my environment, but others I welcome, because I have infinite faith that the future I head towards will be different, improved and definitely worth it. Call me naïve if you want.

My Grandpa's Socks

February 8, 2017

Jack Penner
Georgetown University School of Medicine
Class of 2018

WHENEVER I GO TO THE HOSPITAL, I wear my grandpa's socks. They looked distinguished on an older man, but a little childish on a me, a 25-year-old medical student. I'm okay with that. Feeling like an overdressed kid on Easter helps to balance the overwhelming pressure of becoming a physician.

I still see Pop sitting in a chair with his silver hair that was too strong to fall out during chemo. He's half smiling, with a slight eyebrow raise. Even just thinking about him, I can smell that nostalgic mixture of moth balls and Polo cologne. While one elbow rests on the armrest, his fingers fiddle with a small scrap of paper he always seemed to have. He has one leg crossed over the other, his pants halfway up his shin to display his timeless socks that match his timeless sweater. They were usually wool, sometimes cashmere, solid or a traditional argyle, in classic shades of navy, maroon and grey. Noticeable, yet subtle, they represent maturity, humility and composure. Pop, like his socks and outfits, always seemed so together. This is the Pop I knew.

Then there's the other Pop. The one I only heard hushed, whispered stories about. This man is belly up, sprawled out on a diving board. Iron weights dangle off his ankles a few inches above the surface of his backyard pool. As the breeze rustles the tree leaves, the winter sun casts a paltry spotlight on a suicidal alcoholic. My grandpa. Pop.

Pop set high expectations for himself. He had to be strong, independent, and successful. He hated the idea of burdening anyone else with his issues, so he swallowed whatever life threw at him. He wanted to be perfect.

With these heavy expectations, Pop, like all of us, had to cope. Opening up or asking for help wasn't an option; it would have exposed the fact that he couldn't handle it, whatever "it" was. At first, drinking eased his self-imposed pressure. Eventually, it yanked him into a self-destructive cycle, rip-

115

ping away the curtain of security that hid his inability to live up to his ivory tower ideals. Unable to meet unrealistic standards, he felt like a failure. Failing, in his eyes, was a waste of life. So, he tried to end his.

Full of pills and alcohol, with weights tied to his legs, he waited on the diving board, hoping to pass out, fall forward, and sink to the bottom of the pool. Luckily, he passed out and flopped backwards.

I don't just wear Pop's socks, I wear his demons, too.

Like him, I fear failure. Like him, I want to be perfect. Like him, when faced with high expectations and left to my own devices, I feel the pull towards isolation and the self-destruction. Not wanting to expose my weaknesses or be a burden, my gut tells me to put on a smile, shove down the uncomfortable emotions, and white knuckle through the hard times.

The last time I talked to Pop, he was nearing the end of a steady, but peaceful decline. After overcoming the suicide attempt, a collapsed lung, pancreatic cancer and a heart attack, his body was finally giving out on the last of his nine lives.

As I sludged my way up to lecture, I tried to keep it together. My hood covered my tear-streaked face, my silence covered my trembling voice. I walked extra slow, savoring one last chance to get one last lesson. Today, he opened his textbook on life to the chapter about *A Bronx Tale*, one of Pop's favorite movies.

"Jack," he said, "the saddest thing in life is wasted talent." My talent, according to Pop, was my potential to become a good doctor. I would waste that talent if I followed in his footsteps, suppressed my internal struggles and walled myself off from support. Those behaviors, he told me, brought him to the diving board that day. Connecting, stepping out of his own head, and opening up to others helped keep him from going back.

Talking to Pop was like getting the answers to a test you didn't know you were going to take.

Now, a year later, as I'm trying to come into my own as a physician, I doubt myself and fear failure everyday. The mental, physical and emotional demands of medicine compound these insecurities in a culture that often refuses to acknowledge they exist. The high-achieving, overly independent atmosphere pushes students to prop up a pristine image of strength, competence and unwavering resilience. I wish it was that clean on the inside.

Amidst the long hours, competing demands and big tests, we struggle to find time to pee, let alone process the inner turmoil that comes with grieving families, dying patients and tracking our own fulfilling path in medicine. On top of that, none of us want to admit we can't handle it. All the good doctors seem to be emotional fortresses. Most of our classmates, too, at least on the outside. No one really talks about the trying times, so they must always be fine.

Not me. What about you? What's the cost of this culture of silence?

No single experience captures the trial by fire that is medical education better than the age old practice of "pimping" — the diarrhea-inducing time when an attending physician pelts you with a series of obscure, "read my mind" questions in front of all of your peers and superiors. The questions rain down until you get one wrong. If you're lucky, it stops there. If you're not (you usually aren't), the pelting continues until you become a babbling mess who can't decide whether to shrink into submission, cry or simply soil yourself. Meanwhile, everyone externally cringes for you and internally wipes the sweat off their forehead because they aren't under the lonely spotlight.

One day after my shift, I was sitting at a coffee table in the hospital, trying to decompress after getting the intellectual noogie of a good pimp. This one came too soon after the death of a young husband and father that I had been caring for. As I obsessive-compulsively bite my nails, my mind jumps between ruminating on the barrage of questions I just got wrong and wondering what more we could have done for that man and his family. I'm rattled, festering in my own head, and feeling like a failure.

"You all right?" my friend asks as she walks up. She looks just as tired and a little less beaten down than I am.

I want to be strong and perfect like I'm supposed to, so I start reply with, "Yeah, I'm fi–." Before I can finish, she cocks her head to the side and gives me a look of unbridled disbelief.

"Oh shut up. Come on." She is unapologetically direct and unrelentingly caring.

I laugh. "It's that obvious, huh?"

As she flops down into the seat next to me, I unload the last three weeks on her. Moments of grief and frustration, along with the wonderful ones that reinforce the reasons we go into medicine in the first place.

She does the same, recounting the patients who touched her life and the veritable storm that opened up during her PowerPoint presentation on "Diagnosis and Treatment of An Itchy Anus." Her supervising physician decided to turn a grammatical error into a personal attack on her attention to detail. "What other mistakes will you miss?" he asked. She, too, feels like a failure.

I make a pitiful attempt at a joke about butts, and even though I swing and miss, it at least helps us laugh off some of the absurd aspects of this whole medical student thing. In our self-organized therapy session, we realize we have the same vulnerabilities, fears and insecurities. We're not alone in this.

I think of the last conversation I had with Pop. This is why I wear his socks. To remind myself to break the silence. To remind myself that we settle our inner turmoil with the support of others. Most importantly, to remind myself perfection is neither a realistic human quality, nor one worth seeking.

"I needed this," she says. I nod in reply.

Three hundred to four hundred physicians kill themselves every year. One in four medical students suffer from depressive symptoms, and it just gets worse in residency. Refusing to ask for help, self-doubt, unrealistic expectations of perfection, and loneliness are at the core of these painful stats. We all agree that something has to change. But medicine is a big ship that takes a long time to turn. I'm not sure we can afford to wait for a top down cultural shift to start the conversations around the fundamental highs and lows of the medical student experience. The conversations that remind us we're not alone.

Every time we choose to swallow the difficult emotions, we waste an opportunity to support ourselves and our colleagues. If we refuse to connect with and understand our own emotions in emotionally trying times, how can expect to connect with our patients'?

As his health deteriorated, Pop loved to discuss with me his echocardiogram results, medication changes, and fluctuating prognoses. "My East Coast Doc!" he'd say when I answered the phone. "As your career goes on, you can think of me and all my woes." He's referring to the heart failure, the pancreatic cancer and the lung he damaged falling off a ladder.

I told him I would always think of him. And I do. Every time I put on his socks and go to the hospital, I think about Pop and all of his problems. Just not the ones he hoped.

Second Day as a Surgery Student

June 1, 2017

Lexy Adams
Penn State Hershey College of Medicine
Class of 2018

"There must be a better way to make a living than this!"

Slam.
Silence, except for the persistent heartbeat.
The beat of the ticking time bomb, the dying heart.
It beat uselessly, against flapping intima, seeping vessels
Blood oozing and clotting and weeping everywhere.
An aortic dissection — a dissected body
Laying open, uselessly repaired.
Hand over hand squeezed dripping laps into the cell saver
For hours
Warm blood waterfalling over my student hands while
Seasoned surgeons grafted against a ripping aorta
A stranger, wandering and confused, memory meandering.
Surgeons thought it was hopeless,
Left no other choice.
Intimal flaps flipping across valves, blood pressure bottoming
To the operating room they went
Middle of the night
For hours
Sutured in a bright new aorta, came off bypass
Every needle puncture gushed. Coagulating, bleeding.
Defeated, surgeon threw down instruments, stomped from the room,
exclaiming.
Slam.

It lingered.
Resident crossed to other side of the table, handed me the needle.
Fresh flesh, best practice.
Heart still beating, close the chest.
Blood still oozing
First time suturing
First patient dying
Under my hands, must approximate edges
Not too big of a bite now
Family waiting, don't butcher him
Nurses impatient, huffing and cleaning
Resident critiquing
Hands shaking, not breathing, patient still dying, patient still bleeding
Flat line.

Don't breathe; don't let them see you cry.
First patient dies
While you piece them back together
You feel the heart stop beating.

On Knowing When to Stop and Breathe

February 3, 2017

Tiffany Lin
David Geffen School of Medicine at UCLA
Class of 2019

W E FINISHED PSYCH BLOCK a few weeks ago: five weeks of learning not only about patients' psychiatric disorders, but also physician mental health — sometimes jokingly referred to as "psychation." It's true — in comparison to other blocks, psychiatry is less overwhelming, mostly because the brain is really difficult to understand. It is even harder to figure out how humans perceive the world in their minds, so there are many mechanisms that have yet to be elucidated. As such, this block is a bit of a break from the normal breakneck pace of medical school.

Mental health has been on my mind lately, but not only because of the "Physician Mental Health" and "Resiliency Training" lectures we've been receiving during this block. A few weeks ago, one of my best friends from home texted me to say one of her medical school classmates had committed suicide. He was a second-year medical student in the middle of his training to become a physician, just like I am. Even all the way across the country, the struggles that medical school students face are much the same, and though I did not know this student at all, his suicide felt like it hit close to home. Only medical students can understand how unique of an experience these four years are, and we form strong bonds as we are challenged to learn, struggle, and grow together during this journey. He could have been any one of my 179 classmates who I see day after day.

After the initial shock wore off, I was not actually as surprised by this news as I would have hoped. Throughout this past year, the heart-wrenching stories of suicides by medical students and physicians have only continued to accumulate. It makes me feel tired. How can members of a profession that claims to bring hope into peoples' lives become so hopeless? Clearly, mental health is equally important to both the physician and the patient. Physicians are incapable of caring for their patients if they themselves are preoccupied by personal issues or are facing burnout. My medical school has made it a point to expand their mental health program to make sure students are able to access the needed services. Unfortunately, when there's a seemingly endless amount of information to learn and absorb, in addition to the countless advocacy groups, research projects and student organizations medical students are involved in, time for self-care is often the first thing to go from our packed schedules.

As a patient, being surrounded by a loving and supporting community was one of the largest blessings in my life. They made it possible for me to make it through that year and a half intact, with hope and with a smile on my face. When I grew tired and weary of the pains in my body, my friends listened to me complain endlessly, then encouraged me to face the next day with a better attitude. When my first surgery failed, my parents never wavered in their determination to seek out all available resources and options. When I felt like everything had been turned upside down, my friends reminded me to just breathe as they tried to help me restore some sense of normalcy to life. Though disease infiltrated so many aspects of my every day, and it was tempting to continually wallow in self-pity, I learned that in order to persevere and survive, I had to actively step outside the bubble of Cushing's disease and stop allowing myself to be defined primarily by my illness. My community never ceased to share my burdens, to rejoice with me in small triumphs and to remind me that I have so many more identifiers than just a patient.

To my slight dismay, being in medical school is honestly not that different: It's a difficult experience, and it can feel overwhelming and consuming. Ironically, as I was hoping to move on from my disease and leave it as a chapter behind me, my experiences as a patient — the coping skills I learned, knowing the importance of the community on which I relied — came in handy as I learned to navigate the ups and downs of medical school.

One of the biggest realizations I had between the first and second years of medical school was that, just like I had to step outside of my Cushing's bubble as a patient, I also had to put boundaries on medical school. I learned to be okay with not studying all the time and not knowing every single detail of every lecture. I am happy to be a medical student, but I am also a daughter, a friend, an avid traveler and so much more. For me, it's far more important to block off some time to visit my family, to catch up with friends and to allow myself mini breaks so that I can be refreshed and have a healthy relationship with medical school.

In both roles, as a patient and as a physician (in training), it makes all the difference to have time and space to process what's going on in other aspects of life, and to have friends, family and classmates who can walk on the path alongside you.

To some extent, the personalities of medical school students make it hard for us to acknowledge that we can't do everything alone and that we may need help. We like to plan things way far in advance, we like to be in control of all situations, we like life to be predictable, we try to figure out and understand everything. We are used to having our lives together all of the time. We commit to 200 percent when, like everyone else, we really only have 100 percent to give.

I don't want to hear any more stories of suicide within this community. So I plead with you: Dear friends, as difficult and stressful and consuming this journey may be, please take care of yourselves — mentally, emotionally, physically and spiritually. I'm sure our future patients, colleagues and family members will thank us for it.

Slow Down, Reflect, Recover

June 24, 2015

Diem Vu
Mayo Clinic School of Medicine
Class of 2016

F OR MOST MEDICAL STUDENTS, the third year of medical school is their introduction to life in the hospital. This results not only in exciting learning opportunities, but also emotional tolls — grief, fear, anxiety, exhaustion — that can lead to serious problems including burnout, depression and anxiety.

In the first week of my clinical rotations, a patient in the psychiatric emergency wing looked me in the eye and told me she wanted to murder someone. A few weeks later during my surgery rotation, I was part of a discussion that involved telling a patient with esophageal rupture that he had approximately 48 hours to live. On my neurology rotation, I attempted a lumbar puncture at the bedside, slowly plunging a frightfully long needle into a woman's back as she moaned in fear and discomfort. During my OB/GYN rotation, I moved an ultrasound probe over the belly of a woman who had noticed bleeding and decreased fetal activity; this pregnancy was highly desired by her and her partner and I remember beads of sweat forming on my face as I searched for the elusive, oceanic sound of a heartbeat. I am sure my classmates have also encountered similarly jarring, stirring or frightening situations in their third year.

Despite all of these events, when asked by friends or family, "How are you?" I occasionally reply with a blank stare or phone silence. One day, I realized why this was happening: I was depersonalizing. In the comfort of my own home, I locked away the emotionally heavy events of the day, distancing myself from them. In a way, trying to be a normal 20-something-year-old coming home from work to relax and watch some Netflix, instead of a future professional in charge of saving lives who had witnessed a grisly acute event on the floor or watched someone be told they had cancer or miscarried. The empty gaze and the delayed response: external manifestations of an internal struggle to open the door to those experiences again.

So whenever this happens, I know I need to do three things: slow down, reflect and recover.

Depersonalization is the development of distorted perceptions of oneself and others, a separation of oneself from emotions that ultimately manifests as a lack of empathy. It is unfortunately a common phenomenon among medical students, starting from their first-year gross anatomy course when many students encounter death for the first time. It is difficult to memorize every single orifice and nerve fiber in the human body, but to do so while encountering mortality and the strong emotional reactions that come with it is an even greater challenge. So, while coping, students inadvertently fall into the trap of locking their emotions away. In general, students depersonalize to protect themselves from being hurt by negative experiences — they objectify patients or dull their emotional reactions to certain situations. It just so happens that depersonalization is one of the aspects of burnout, a syndrome also characterized by dysphoria and emotional exhaustion. The emotional burdens of clinical clerkships are also enormous and many students find that burnout gets worse in their third year.

For this reason, writing has been personally important for me throughout medical school and especially so in my third year. Writing offers reflective opportunities that allow students to process powerful emotions or struggles, to slow down and face the emotions that they may have suppressed in the process of coping with the demands of medical school.

For me, this has meant jotting down a few lines on my service lists when I wasn't too busy, sitting down to write a simple haiku or scribbling the events of the day in a journal or planner. These exercises helped me unlock and reclaim my emotions. They helped me recover from exhaustion and depersonalization. And looking back on them, these scraps of writing tell the story of my third year. I have faith that continuing to create more of them will help stave off cynicism and burnout. They will someday tell the story of my transformation into a physician.

There are many interesting papers concerning the effects of writing and reflection on health care providers. In a recent randomized clinical trial, journaling, reflection and group discussion helped physicians experience significant improvements in meaning, empowerment and engagement in work. A 2013 survey revealed that the majority of medical students at one institution found that reading a medical school literary publication promoted patient-centered care, prevented burnout and helped them understand their colleagues and classmates. But when I mention writing to my classmates there tends to be a sigh, followed by a declaration that runs along the lines of: "But I'm an awful writer!"

We all have experiences, and experiences can be captured in writing by anyone, regardless of their publication history, their creative writing training, or their knowledge of literature. I firmly believe that it doesn't matter how much Whitman a person can quote or how many local coffee-shop readings a person has ever been to: Everyone has a story to tell.

As I shared at the most recent "Examined Life" conference at the Roy J. and Lucille A. Carver College of Medicine in Iowa City, I discovered that there are many writing exercises that can not only help get a poem started, but also document and dissect a medical student's experience. For instance, simply jotting down a to-do list or dissecting the contents of a briefcase or backpack can lead to the creation of a catalog poem that captures a snapshot of medical school life. Spending a few minutes recalling the events of a day in the form of a journal entry, or writing about a patient encounter from the perspective of an object in the room, can help a student relive an experience through new eyes, rediscover emotional responses, or find opportunities for change.

These exercises only take a few minutes, and anyone can do them. All you need is to set a timer for 5, 10, 20 minutes and to resolve not to lift your pen from your paper during that time. Just write. Not for publication. Not to impress anyone. Write for yourself.

My school has a third year extra-clinical curriculum called Safe Harbor, which unites the class from our varied rotations for a short discussion session concerning topics like bias, conflicts of interest, professionalism and work-life balance. It has been an excellent forum for reflection, a resource that I know students at other medical schools do not have. At the "Examined Life" conference, I watched presentations concerning the development of reflective curricula that were similar to Safe Harbor. These curricula are new and in high demand among the students at other schools. This made me grateful to have a safe space to reflect, a place to reconnect with my friends again and listen to their stories.

But I would also advocate individual reflection through writing — or any other activity, like prayer, meditation, exercise, music, or art — as a means for students to slow down, stay balanced and stay emotionally intact throughout the challenges of medical school.

We find ourselves in some terrible situations in third year, but we also get to have the most amazing experiences: birth, healing, hope. And, even the worst experiences are slices of patients' lives that we should consider ourselves privileged to encounter.

Because of that, when someone asks me how my day was in clinic or in the hospital, I want to say something more along the lines of, "It was incredible! Let me tell you..." And so I try to slow down, reflect and recover as often as I can.

In some ways, the fourth year of medical school feels like the last hoop to jump through. Students are constantly reminded that residency can be brutal, that this is the final time to get education without true responsibility, and that loan repayments are just around the corner and yet — it can be hard to enjoy the final year when there's so much to get in order. This section is aimed at addressing some of the challenges of fourth year with practical advice from our now-alumni. Jarna Shah, Ajay Koti, and Farah Khan describe the process of deciding on a specialty — a choice that is effortless for some and riddled with doubt for others. Eric Cotter, Lisa Cotter, and Michael Cloney reflect on trials of the match and interview season, a time that is bizarre and exhausting both for applicants and their loved ones. Our "Match Day Spotlight" series is an annual favorite, celebrating members of the *in-Training* community who have made it through medical school's many milestones and are about to find homes in their respective specialties. We hope it is an enjoyable read for students and their team alike.

The Final Year

Testimony by Vanessa Velez
Class of 2018 at University of Texas Medical Branch

The Final Year:
Selecting a Specialty

(Foolproof) Guide to Choosing a Specialty

March 25, 2016

Jarna Shah, MD
University of Illinois at Chicago
Class of 2019 Resident Physician in Anesthesiology

E VERY MEDICAL STUDENT dreams of having that "Aha!" moment where you instinctively realize your future specialty. Unfortunately, it never seems to be so simple. That moment is often insidious and occasionally tainted with self-doubt.

That being said, here is advice on choosing a specialty, based on experience: that of my classmates, physicians, residents, faculty, family and friends.

1. Don't expect to know your specialty coming into medical school.

Some students enter medical school with a clear calling on what they wish to pursue. It's perfectly okay to have that direction, but it's also okay if you don't. Do not be afraid to say, "I don't know." You are not required to know your specialty choice up front — but if you're one of the lucky few who always had a clear vision on your specialty, kudos!

Be aware of your personal biases and try to avoid limiting yourself to a certain field. You will quickly find that your experiences as an undergraduate are innately different from working as a medical student on the wards. The whole objective of medical school is to have new experiences that will shape you as a future physician. Embrace the uncertainty. Don't fear the unknown.

2. Think about what you enjoy doing.

It is time to start seriously considering what you like and dislike about the many facets of medicine. Looking at the potential list of dozens of medical specialties, it's very easy to be intimidated.

What aspects of medicine do you enjoy most? Do you like working with patients in the inpatient or outpatient setting? What type of patients? Young, old, sick, healthy? Do you prefer to work from a distance? What kind of relationship do you want to have with your patients?

Do you see yourself as working mainly in the operating rooms? Do you like using your hands more, your mind more, or a balance between the two? Consider if you like being in the hospital, the clinic, or prefer a mix of both settings.

What about procedures? Do you like fast-changing environments and patient conditions? Or do you prefer to have lower acuity patients with long-term conditions and illnesses? Do you like being a patient's first contact when they are admitted to a hospital, or do you prefer to know every answer in a very specific manner?

More importantly, where do you see yourself in ten years?

Your answers will change as you have new experiences. Don't expect your early impressions as a first year to remain set in stone throughout your years of training. Exposure will mold your interests in unexpected ways.

3. Explore career choices early.

This is perhaps one of the biggest hurdles for a medical student to face. Not only will you be bombarded with exams, assignments and deadlines, but you will be urged to explore specialties and make your decision as early as possible. Personally, I found this utterly terrifying.

You see all those emails for specialty interest group talks? Go to them. Join clubs and explore fields — even if you don't know for certain whether you would enjoy that specialty. Pursue extracurricular activities that interest you. Avoid stretching yourself too thin; do not over commit. Care about what you do. It will show — even if you don't end up going into that specialty.

Talk to your advisor and make sure you have a healthy balance of school and non-curricular related activities. Medical school is strenuous, and you have to look out for yourself and your fellow classmates. Don't forget to prioritize your family and those who matter to you. Medical school burnout is a real phenomenon, and we as future physicians can be pretty awful about prioritizing our own health.

4. Find your niche.

It is one thing to choose a specialty, it is quite another to find your career. Many people don't realize that your future will involve a lot more than just being a physician in your field.

Sign up for mentoring opportunities that your school may offer for first and second year students. This can be through a formal program, or through interest groups providing shadowing opportunities with attendings. If you are unsure where to start, start by meeting with your advisor. Talk to your school's clerkship director for your rotation of interest. Email senior residents and ask if you can spend an afternoon working with them. Talk to your classmates and see how they are getting exposure. Many schools offer longitudinal courses that involve early hospital experiences. Talk to your preceptors and professors — they are fantastic mentors. Even if they are not in the exact field you want to explore, they will probably know whom you should contact. I do not say this lightly, but this step requires effort and persistence. You need to be proactive if you want the opportunity to explore some specialties.

5. Do you like research? Pursue it. Don't like it? Don't do it.

Depending on your eventual field, your research experience may or may not play an important role in your residency match. The types of research opportunities available are immense: bench work in a lab, clinical trials, studies in the humanities or social sciences, global health and health care management. Find out what types of research are available around your campus, and whether you think you would find it to be a worthwhile pursuit for you.

If you do decide to pursue research, here is my advice: show initiative, be attentive and act professional. Ask questions. Do your work. If you find yourself dreading your time in the lab, consider your other options. Life is too short to do things you dislike. Recognize this early.

6. Do some serious soul searching.

This is probably the last thing you want to hear, but it is very important. You need to make sure you choose a specialty that best suits your personality. If you plan to practice for several decades, it should be a career that you believe you can honestly love, and not one you're pursuing for secondary gain.

Each field has a slightly different environment. Take a close look at what personalities are attracted to this specialty and see if you can see yourself getting along with those kind of physicians. Look for a field that aligns with your personal characteristics. I recommend the Careers in Medicine website. Keep in mind that your answers might change as you progress in your training.

You need to be realistic and start thinking about how you wish to see yourself ten or 20 years down the road. You need to know what you're looking for in the next ten years. You need to know the good and the bad about your field choice. Don't neglect your relationships and the things that matter most to you while you make your career choice. Evaluate work life balance. This is important regardless of gender.

7. Enter clerkships with an open mind.

Be enthusiastic about every learning experience, even the ones that are rough. As a third year, it is often easier to think about ruling out specialties rather than finding the one that is best for you. Try not to let that elimination principle stand in the way of experiencing each rotation to its fullest.

Don't discount a field at first glance based on what others say. You are wise enough to make your own deductions. Take time and analyze what you like. Work with other medical students and help each other out. A happy working environment makes rotations fly by and makes them more memorable.

Don't taint a rotation with negative thoughts or complaints. Work hard and keep going. Third year is a powerful experience, even if you find that you don't plan to pursue any of those specialties. Have fun, because this may be your only chance to deliver a baby, dissect an eyeball, or scrub in on a liver transplant.

8. Talk.

Can't figure out what specialty you want to pursue? Talk it out. Call your mother and have a heart-to-heart conversation while you do your laundry. Use your non-medical best friend as a sounding board and walk through your thought process. Having to explain why you like what you do is a great way to uncover what really draws you to the field, and a new perspective is always useful. Talk to your classmates, your upperclassmen and your advisor. It may feel awkward to talk about your feelings, but suck it up and just do it. Trust me.

Talk to residents and physicians, and ask what drew them to their field. I have received amazing advice and perspectives from the residents around me. They are the best source to tell you exactly what they like and dislike about their career choices. I like to ask every resident I work with about what attracted them to their field. I also try to ask them what they consider to be the worst part of their job. A resident once told me a valued piece of advice: When you think about a career, think about the most routine "bread and butter" case that you would experience. Do you think you can take care of that type of patient day after day without getting tired?

M4's are another great resource. They are very honest and open about their own experiences. What better way to choose your field than to see how your peers made their decisions? Check out these M4 reflections from internal medicine, family medicine, psychiatry, surgery and obstetrics.

9. It's okay to change your mind.

If you find yourself heading towards a career and you realize that you are having a hard time committing, be open and willing to look into other fields. It is better you discover this now, rather than three years into your residency in the wrong field. Meet with your advisor. Talk to your school's program director. Lay everything out and ask their opinion. You are not obliged to follow everything they say, but they are the best people to ask and will provide you with the most definitive answers.

10. Don't panic.

This whole process is not an easy decision. Remember that your peers are going through the same process as you, and that you are not alone. If it gets to be a bit much, take a step away from the decision process, spend time doing things you love with people you care about, and then return to tackle the mighty beast. Don't forget why you went into medicine in the first place.

How will I know I've made the right choice?

You will find what you love. Right now, everything may feel like an up-hill challenge, but do know that you will emerge from this experience as a stronger and more confident individual. Realize that your love for your field will increase as you become more competent. Choose something you think you would truly enjoy doing.

Are you happy when you are in that field? When I found my specialty, eight hours would fly by and I'd still be rearing to go. Going to work didn't feel like a chore. I missed being there on my days off.

It's okay if the light bulb doesn't suddenly flash on the first day you spend in your field. It's okay to bump gently into your field. You're in for a great adventure. Good luck!

A Third Year Opus:
Curtains and Red Flags

December 14, 2016

Ajay Koti
Morsani College of Medicine at the University of South Florida
Class of 2017

Part One

I was quite confident that it could happen to other people. I was quite confident it wouldn't happen to me.

What do you want to be when you grow up? I mistakenly expected this question to be laid to rest after starting medical school. But it comes to occupy a position of great priority and urgency: *Sure, you want to be a doctor, but what kind?* — as if the former decision had been a minor one.

The considerations in choosing a specialty are multiple. There are matters of lifestyle and compensation, of competitiveness and rigor. There is the push-and-pull of breadth versus depth, of procedure versus prose.

Fortunately, medical students are surrounded by people who can help navigate this existential mire. There are more senior students who, leaning back and angling their chair legs off the ground, offer: *The three things you need to figure out are [platitude, platitude, platitude]*. There are online assessments — powered by reliable, rational algorithms — that give different results based on the time of day and day of the week. *What changed — the test or the test-taker?* There are residents and attending physicians who will either share their own journeys or bark: *dermatology!*

At first, the indecision is thrilling. It makes available to the imagination limitless possibility. But as Time pushes on, the indecision becomes inconvenient, then vexing, then torturous. Because, at some point, pros need to outweigh cons and plans need to be made. *Surgery?* needs to become *Surgery, period*. Decisions must be made.

I counted myself fortunate that I was spared from grappling with this process. After a few months of obligatory vacillating and non-committal statements, I quickly and emphatically decided on family medicine.

My resolve on this was ironclad. By my estimation, there were only two significant drawbacks. First was the prospect of joining the ranks of one of the most beleaguered medical specialties. I was not blind to the statistics on primary care shortages, the burdensome scale of patient panels or the grind of squeezing thirty minutes of patient care into half that time. Far from being repellants, these challenges were irresistible to my sociopolitical (read: bleeding-heart) inclinations. *You got into this thing to serve others. It is thus entirely appropriate to do so under relatively punishing conditions.* I would make the requisite sacrifices in service of medicine's moral mission.

My messiah complex is a subject for entirely another essay.

The second problem was less assuredly dispensed with. The broad scope of family practice also meant a broad patient population — one that included children. Hmm. For reasons that I could not articulate, this produced some hesitation. Children. Hmm. Kids. Youths. Yoots.

I tabled this concern, deciding that I could address it later. At any rate, it made no difference in how I felt. When people inquired about my plans, family medicine was always the answer. Those who had been through the process told me to keep an open mind during clinical years, that they started medical school so certain of their specialty, only to fall in love with something completely different. I nodded politely but hubristically dismissed them.

I was quite confident that it could happen to other people. I was quite confident it wouldn't happen to me.

Six weeks of pediatrics zoomed by, and I could feel myself falling in love with it. The diversity of ages. The role of the parents and extended family. The opportunities for education and advocacy. The boundless resilience of body, mind and spirit that is revealed when a sick child becomes well.

Pediatric patients harbor little of the cynicism and tunnel-vision that plagues adult life. We are all afforded exposure to new experiences, ideas and emotions every day, but adults are less sensitive to them. Children feel the exhilaration of novelty in their bones. A 16-week-old smiles broadly at a new face, a sixteen-month-old marvels at the occupants of a fish tank, a sixteen-year-old forgets his iPhone and his angst, even if only for a moment, to play with a therapy dog.

To experience life through the eyes of a child. That was, and is, the promise of pediatrics. You might imagine my irritation at being reduced to a cliché worthy of the Lifetime channel — an heretofore narrow-minded medical student discovers enlightenment and broader horizons — but the unqualified enjoyment I felt on the pediatric floor was ample distraction.

In retrospect, my initial resistance to the specialty was rooted in ignorance. I had spent little time around children, and I was, frankly, intimidated by them. Adults were so much more straightforward — you could talk to them. They could tell you what they needed from you. How on earth do you talk to a child?

The answer is simple, of course: You just talk to them.

But when the patient is just one week old, the conversation is a little one-sided.

Part Two

Baby Ahmed was swaddled in a crib. His father sat in a chair adjacent. His mother sat behind a curtain. His sister, three years his senior, bounded about the room, indefatigable. Ahmed, by contrast, did everything one expects a nine-day-old to do: not much. For most of his five day admission, he slept, stirring only to feed or yawn, his mouth gaping to gargantuan proportions relative to the rest of his jaundiced body. Behind his shut eyelids was more evidence of the jaundice and feeding failure that caused a 13 percent weight loss and a direct admission to a hospital room for failure to thrive. The hospital machinery whirs to life — fluids, labs and lights.

Medically speaking, failure to thrive in a child is not a particularly esoteric diagnosis. Its management is similarly straightforward, premised on rehydration and addressing the underlying cause of the weight loss. A lack of education? A lack of care? In Ahmed's case, it was the former. During the admission history and physical, his father, a generally pleasant man in a stained work uniform, was unfazed by the events. *Ahmed isn't feeding well, so, logically, he must not be hungry.*

His mother, dressed in a traditional Punjabi suit, her face framed by a hijab, sat silent on the other side of a curtain that divided the room. His sister grew increasingly gleeful at the parade of short and long white coats who came to greet them. A face on an iPad translated our recommendations into their native language. Ahmed's mother nodded placidly in agreement behind the curtain, Ahmed's father thanked us and left. (He was due at his second job). Ahmed's sister giggled.

Part Three

Much is made of red flag symptoms in medical school, and appropriately so. These symptoms are indicators to the clinician that the diagnosis is serious — debilitating, or even life-threatening — and demands immediate evaluation and treatment. Red flags of cauda equina syndrome, in which the nerve roots that trail off the spinal cord are compromised, include severe back pain, progressive neurologic dysfunction and both urinary retention and incontinence. As with all things in medicine, it is one thing to learn these from a textbook and quite another to learn them from a patient.

So when our team was alerted of a direct admission from an outpatient site — a six-year-old male with all of the above symptoms — I must admit I did not appreciate the grim faces of the attending and residents.

Comparing Hunter to a pretzel may sound trivial, but I can think of no more apt description. The boy sat contorted on the bed, tearfully shouting away any would-be examiner. His head was in a torticollis-like position.

His parents weren't much older than me and seemed to have no more than a high school education. They explained the four week chronology of the symptoms — beginning with pain that built to excruciation and progressing to virtual paraplegia, all in the setting of fever and weight loss. In that time, they had sought medical evaluation at two emergency rooms, only to be reassured and discharged. Red flags must have been absent or unrecognized — or perhaps the parents, young and ineloquent as they were, were dismissed by a system that is more easily navigated by the privileged. Hunter went for an MRI, STAT.

Part Four

By the second morning, breastfeeding was still not going well. The baby's weight had dropped again, albeit by less. *He's still not hungry*, translated the face on the iPad, as Ahmed's mother looked on expectantly. *She still doesn't get it*, I thought. Ahmed's sister was oblivious; she tried to engage me in a game of peek-a-boo.

Stupidly, I spoke louder and slower, as if this would make my English any less foreign. Perhaps the interpreter translated my irritation as well. Over the day, nurses, consultants, and members of our team visited her again and again and again (much to her daughter's delight), to reinforce the teaching. Each time, she indicated that she understood and had no questions. Each time, my voice dripped with increasing condescension. The thinness of my commitment to patience, to progressive values and to cultural sensitivity was seemingly laid bare. Empty words, easily derailed.

Highlighting the cruelty was her consistently courteous and deferential nature. If only she had been resistant, argumentative, unreasonable — anything that might justify my arrogance. But she seemed to pay it no mind. I wore a white coat and I was a man — it was to be expected. It was the appropriate order of things.

Part Five

The MRI images were uploaded piecemeal just after signing out to the night shift. An enormous tumor had grown in Hunter's retroperitoneum, extending to — and seemingly into — his spinal cord. Hunter had been transferred to the post-anesthesia care unit (PACU), still under the sedation from his MRI. Neurosurgery, pediatric oncology, and anesthesiology were called. Neurosurgery recommended immediate laminectomy to relieve the cord compression; they would meet us and the parents in the PACU, just yards away from the operating rooms where a suite was being prepared. Before that, there was the matter of telling the parents. They would be the last to find out.

The sight of a human being crumpling is not one easily forgotten. Her hand moves, not to her chest — she is not yet heartbroken — but to her abdomen, where she carried Hunter for nine months. Her eyes grow distant and wide all at once, and she chokes and gasps. Then the floor opens up, and gravity pulls her down. She falls to her knees. She crumples like paper and weeps.

We walked the parents down to the PACU, where a surgeon rapidly earned their trust and consent. Hunter remained sedated on a stretcher, his brow dotted by his parents' tears. Then he went into surgery.

Part Six

By the third morning, Ahmed's weight had finally ticked up, and I went to offer words of encouragement. In their hospital room, Ahmed's sister beamed at my entry and tugged at my pant leg. She and her mother were wearing the same garments today that they wore yesterday and on the day prior. On the table, a breakfast tray was largely untouched. The eggs and pancakes were exotic, the bacon was a religious taboo. Instead, they had subsisted from a foil wrapped plate of more familiar foods, brought in by her husband during his brief visit between jobs. Ahmed's mother and sister had not left this hospital room in three days. Beyond the door were unfamiliar white faces, who spoke unfamiliar words, who ate unfamiliar foods, who wore unfamiliar clothes, who lived unfamiliar lives in an unfamiliar world.

The weight of my own ignorance and blindness sank in my gut. How little I knew about this woman, even though she was, for all intents and purposes, our real patient. I had waited three days to learn that she was even younger than I was, that she had been in the United States for just a few months, that she had married at 18 to a man more than a decade her senior. I had waited three days to learn her three-year-old daughter's name (Maya). I wanted to apologize for my impatience, but the words felt feeble and inadequate. I offered to take her daughter to the playroom, so that she might have a break from the toddler's relentless energy.

Half an hour later, little Maya returned to her mother triumphant, regaling her with what I can only assume was a retelling of her voyage onto the unit — the faces of nurses and doctors, with whom she exchanged waves and high-fives, the vibrant colors of an enormous fish tank, the room with a seemingly endless supply of toys.

For the first time in three days, Ahmed's mother smiled.

A year later, I met Hunter again as an outpatient. His tumor responded dramatically to chemotherapy, shrinking enough to make surgical excision possible. Miraculously, he had no lasting neurologic deficits, and he nimbly and idly rolled about on the exam table, eager to leave the clinic and get back to his summer vacation.

I hadn't seen him since the day after his surgery, in the pediatric intensive care unit, where he stared blankly and silently ahead. Now, he looked right at me and grinned. Life blazed in his eyes. I asked what his plans were.

I'm going fishing.

What I Wasn't Meant To Do in Medical School

June 20, 2014

Farah Khan
Ross University School of Medicine
Class of 2014

MEDICAL SCHOOL HAS BEEN weird.
I learned things about myself that I didn't like and I've also been pleasantly surprised by things that I thought I would never like.

As a fourth-year student, your focus and energy are consumed with thoughts revolving around future residency: the labor intensive training that follows four long years of medical school.

What no one tells you is that you will question your commitment to medicine on more than one occasion. Actually, that's not entirely true. Physicians who have been practicing for many years will tell you about the "good ol' days" and in their endless wisdom they will advise you to find a different career, but as a young 20-something-year-old, what are the chances you will listen? You always think it will be different because you have worked so hard to get into medical school and because you were personally affected by so-and-so's sickness and because you want to help people.

You're determined and motivated and energetic. You want to save lives.

Unfortunately, every medical student has those days when no amount of determination, motivation or energy will help him or her get through the painful studying for board exams or help function as a semi-normal human being on two hours of sleep. For some of us, we have more of those days than we like to admit which makes fourth year — with all of the life-changing decisions revolving around which specialty to pursue — all the more interesting.

While it's fairly common to find students who know what specialty they will pursue before they even begin medical school, it's also quite common to find students who change their minds several times and even students who end up in specialties they would never have imagined they would like.

Enter clinical rotations.

The point of these clinical rotations is to expose students to different specialties and different disease processes in various settings. In an age when medicine has become so specialized, it makes sense to experience these rotations. Even if you have no interest in pursuing surgery, you better believe you're going to suck up those tears, strap on your no-nonsense attitude and fulfill those 12 credit hours so you can graduate. Along the way, you may even be begrudgingly surprised at how much you learn.

The weirdest part of medical school for me has been the change I've seen in myself. In some ways, I'm smarter and more equipped to handle life's curveballs. I'm more confident and deliberate in the decisions I make — both with patient care and in my small personal life. That is, until I had to fill out ERAS, the mother of all applications.

When it came time to actually apply to residency programs, I was the most lost I had felt in years. I have a heavy emergency medicine background (I was a medical scribe at a level one trauma center in Northern Virginia for a few years before starting medical school and it's the only thing I have lived and breathed for) and before beginning rotations it's the only thought I was consumed with. How do I get in to an emergency medicine program? Should I choose a three-year or a four-year program? Which hospital should I do an audition rotation at? And in so many ways, my personality was suited for the high energy, think-on-your-feet demanding shift work.

I'd like to fast forward to the part of medical school when I did my emergency medicine rotation and realized that things aren't greener on other side and just like that, I no longer had any interest in pursuing the specialty which had driven me to go into medicine to begin with.

For me, the decision didn't come easy and in a lot of ways it didn't make sense to anyone else in my life, including my parents, friends I had grown up with, friends I had made through school, or attendings I had worked with prior to medical school. Every time I told them I was interviewing for pediatric residency programs, I was met with furrowed eyebrows, an awkward pause followed by a "Hmm. That's ... interesting."

Was it interesting? Sure, it was odd to go from craving a high-stress, fast-paced environment like the emergency room to a mellower inpatient pediatric ward, but the transition had been so gradual in my head that I didn't even realize it was happening. The change seemed so abrupt to the people in my life, but to me, it was the only thing that made sense. I was still a little high-strung and had OCD tendencies, but I was also calmer and more secure with who I was. And somewhere along the way, I had realized that having a personal life did matter to me. It should be okay to change your priorities as you get older. I thought that was normal. But then again, I've been in medical school for the past four years, so I probably don't have a great grasp on what normal even means anymore. My normal has been studying in the library for 14 hours on a Saturday, working overnights on holidays, and living out of a suitcase for the past year and a half while traveling for clinical rotations in different cities.

The surprising response to my choice in specialty weighed on me more than I like to admit. I thought I was more confident and an adult who could make more deliberate conscientious decisions. Or at least be able to handle others' skepticism a little better. I was trying to reconcile who I thought I was going to be and what I thought I was going to do with my career with what I actually ended up enjoying in medical school.

As the deadline loomed near to submit our final rank lists, I had more and more tough days when I wrestled with why I even applied to pediatrics and why I had been pushed away from emergency medicine. Because I like working with kids. Because I like having an impact on a teenager's life. Because I like the flexibility in what kind of medicine I can ultimately practice. Because I felt like I was making a difference. Because working with kids makes me happy. Because, well, because it made sense to me.

While I know my friends and family have been slightly confused by my choice, it's also been weird for me to try to process their reactions and to maintain a rational thought process in my head. Medical school is a strange journey that we voluntarily inflict upon ourselves. It challenges you in more ways than one. It pushes you to be the best you can be while simultaneously making you beg for mercy.

For those of you coming up on the Match: Good luck and don't let the transformation you see in yourself scare you.

The Final Year:
The Match

Planning for the Couple's Match: Our Experience

May 31, 2017

Eric Cotter and Lisa Cotter
Georgetown University School of Medicine
Class of 2018

M ATCHING INTO A RESIDENCY program is the culmination of four (or more) years of incredibly hard work and determination. This process does not come without an abundance of stress, fear and at times, self-doubt, at least in our experience. Those who are seeking to couple's match face a magnified challenge in that they need to not only impress a residency program in a specific city or region, but have their significant other do the same, ideally in the same city or region. We are facing the couple's match in 2018 and would like to share the advice we have received from our advisors, professors, and friends in medicine, as well as our own thoughts on the couple's match and planning away rotations.

Limited Data

One of the hardest aspects of attempting to couple's match is the paucity of data that exists on the topic. For those unfamiliar with the process, the National Residency Match Program (NRMP) allows any two applicants to "couple" by linking their residency program rank lists so they can match to a desired pair of programs (often done for location purposes). The NRMP publishes an abundance of data each year, breaking down the nitty gritty details of the match for that year, complete with breakdown by specialty. However, we have found a striking lack of analysis of students who have declared that they are couples matching. The only statistic available from the NRMP is that a record-high of 1,046 couples participated in the 2016 Match, 11 more than in 2015. Furthermore, 95.7 percent of those declared as couples matching were matched to first-year positions, the highest on record with the NRMP.

Advice for the Couple's Match

As one might expect, the limited information makes the process of couple's matching all the more anxiety-provoking, especially in situations where both individuals desire a highly competitive specialty. We have had many open and honest conversations with each other before seeking specific advice from our advisors and we knew that in order to successfully make it through this process, we would revisit these conversations again and again ... and again. This thought mirrored the best piece of advice we received. It came from Dr. Mitchell, the marvelous dean of our medical school, and it was to simply talk about everything. While we would never dream of asking one another to put their dreams on hold for the other, we know couple's matching is not without compromise. Discussions regarding residency should be a fluid process. When you decide to couple's match, your dreams are linked. We have been wholeheartedly supportive of each other's dreams from the start in each of our pursuits of a medical specialty.

Compromises and Sacrifices

There will be a point, both when deciding on away rotations and when ranking programs, when both individuals need to assess what is more important to them: getting into a specific program or going to a certain geographical location. Reasons for the former include prestige of the residency, quality of training or fellowship connections. Reasons for the latter include closer proximity to friends and family and a higher concentration of programs in a location that can increase chances of being together. To add a layer of complication to the mix, different institutions are often better known for specific medical specialties. Therefore, what might be a dream program for one person may be a less desirable option for the other, necessitating discussions regarding what each person is willing to sacrifice. For away rotations, we were able to identify programs each of us liked from a training standpoint that also have additional programs in close proximity. Ideally, this will help facilitate our ultimate goal of living together for residency. For other couples, it may be more important for one person to get into a specific program, even if that means a further commute to the other partner's residency location. There is no right or wrong. Each couple has different goals going into the couple's match; you just have to find what works best for your situation and make it a frequent point of conversation. Establishing goals through tough conversations early on will help ease some of the tension over decision-making later on.

Be Realistic About Your Applications

We would be remiss if we did not state for those who are couple's matching that you have to be realistic about your application and how the two of you look together on paper. Board scores, grades (particularly in the clinical years), honors and awards, letters of recommendation and extracurricular activities matter every bit as much as if you were applying as an individual. However, when couple's matching, both parties need to carefully consider the strength of their partner's application as well. If one partner is not a strong fit at that institution, then no matter the strength of the other, alternative plans might be in order.

Geographic Considerations

Another piece of advice we received was to choose away rotations in cities with at least two programs, but preferably more. This is sound advice, as large cities like New York or Chicago offer five or more programs in nearly every specialty, allowing a multitude of ranks within the same city (albeit not necessarily at the same institution). In addition, from speaking with our advisors and medical colleagues, if one partner is offered an interview at a program while the other is not, it is 100 percent acceptable to kindly call the program or have program directors reach out. Letting them know that your significant other is interviewing there and that you have a strong interest in the program as well is one way to gain interviews at programs that may otherwise pass over your application. A gentle, respectful nudge might go a long way in terms of gaining interviews, as many programs value the happiness of their residents. We have several friends who are currently in the process of couple's matching and they have stated that many programs (in a diverse range of specialties) are very receptive and are used to receiving these requests. A longer rank list only betters the chance to not only live together, but also to match into your desired specialty.

Conclusion

While no one at our institution has much, if any, experience specifically advising those trying to couple's match into dermatology and orthopaedic surgery, the advice we have gotten has led us to this: We are going to do everything we can to maximize the number of ranks in the same city (or neighboring city) in order to be as close to each other as possible during residency, even if that means not matching into our "dream" programs. While the coming months are sure to be stressful and test our relationship like never before, we have faith that our love and support for each other will endure and we will both come out smiling on Match Day 2018.

How Your Surgeons Got Their First Jobs

June 22, 2016

Michael Cloney
Columbia University's College of Physicians & Surgeons
Class of 2016

T HEY HANDED ME AN out-of-tune Martin, a high-end piece of acoustic craftsmanship, because I'd listed "electric guitar" as a hobby on my résumé, but there was no amplifier in the room where they were interviewing me. I tuned it by ear as best I could while the four of them watched me silently in their suits. Then, with a nickel from my pocket in lieu of a pick, I tried to muster the gods of heavy metal with a guitar meant for bluegrass or blues. When my one-man, acoustic rendition of Metallica's "Master of Puppets" had persevered for longer than was comfortable, they were satisfied.

"That was terrible, but you pass," said the grayest among them. They chuckled, and I chuckled. Now, I had to convince them that I should be a neurosurgeon.

After four tough years of medical school and student loans, facing down seven more grueling years of training, aspiring neurosurgeons enter a residency interview process that is as bizarre as it is effective. Your residency is your first job. It's where you learn your craft and launch your career. You'll spend as many years at your residency program as in middle school and high school combined, and more than in the average marriage. You'll share around 88 stress-filled hours per week with an intimately small group of colleagues, who you'll see more than your friends, family or the light of day.

I hadn't been playing guitar much recently, as I was traveling, and busy, and busy traveling, sleeping on friends' couches or renting cheap hotel rooms with my student loan money. But I always stayed as close as possible to City X's major academic medical center, because I couldn't risk being late for my interviews in the morning.

And those mornings could be hellish. I would usually make my way from my hotel to the interview at around 7 a.m., tackling my unfamiliarity with the city I was visiting while grappling with sleep deprivation and a hangover, and recovering from having swallowed steaks to the point of discomfort only hours before. Then, I would explain why I should be the one operating on their patients' brains.

This particular interview day started typically. I awoke to the sudden and stinging sounds of smooth jazz from the hotel alarm clock, evoking a throbbing pa-pain-pa-pain-pa-pain between my temples. The previous night had ended five hours prior, and I felt terrible. My dimly lit hotel room seemed blindingly bright, piercing straight through my pupils to my pituitary. I didn't leave myself enough time to dawdle, so I rushed to throw on the suit I'd recently bought online — "modern fit" and "charcoal grey," perfect "for the gentleman on the go!" I let out a quick sigh and exited the hotel room, starting the act of being as awake and professional as I could.

My hangover resulted from a crucial part of the process. This interview, like most, was a two-day affair. One day had a social agenda, the other a professional one. Too few hours before my embarrassing acoustic Metallica cover, I was at the all-important social dinner, where applicants broke bread with the residents of the program. No faculty present. If recruited, we would spend most of our time alongside these residents for the next seven years. We watched how they interacted with each other, carefully considering whether we could see ourselves among them. They, in turn, would size us up, seeing if we would fit in as part of their crew.

As was typical, the dinner was held at a steakhouse that no one in the group could afford. The neurosurgery department picked up the tab using a "recruitment budget" that planned for three or four such interview dinners per year. The residents were in high spirits — interviews meant a light schedule tomorrow, with no surgeries on the queue. They could fill their bellies on the department's dime, with food and drinks that easily beat the hospital cafeteria fare they were used to. We would all be having the tasting menu that night, and the wine bottles outnumbered the diners. I later looked up "tasting menu."

Behind the seemingly laid-back atmosphere is the competitive pressure of The Match, which can bring out the worst in some people. One study found that more than 30 percent of students applying to neurosurgery may be misrepresenting their accomplishments on their résumé. Medical students tend to have plenty of résumé-type accomplishments, but some begin to think that every line on their application is make-or-break and that blank space is a liability. Two years ago, a student I knew was expelled for exaggerating his application, only months away from graduation and with hundreds of thousands of dollars in debt. Worse still, he was a stellar candidate who would have done very well if he had only been honest.

But in medicine, honesty is as essential as embellishment is dangerous. Health care is delivered in teams, and team members have to know that they can rely on one another. Physicians are a proud bunch, and neurosurgeons are definitely no exception. But, it is dangerous for a surgeon-in-training, lacking in experience, to also be without the candidness to say, "I need help," or "I don't know." Hiding weaknesses on a résumé foreshadows hiding weaknesses elsewhere, and calls into question the accuracy of what you say about the latest brain scans for Mrs. Jones in bed 209.

So when four senior neurosurgery professors asked me to play something for them on the guitar, what they were really asking me was, "Are you a liar?" And, as I scraped the nickel on Jefferson's face across the wound strings on the guitar's lower register, I was really responding, "No, sir! Absolutely not, sir!" even while my lips were saying, "Sorry — 'Master of Puppets' sounds better with distortion."

Now, in my own defense, I would have played better with more hours of sleep or less to drink the evening before. And while that wine-coaxed personality test served an important purpose, it's not hard to see how such nights can get out of hand fast. Take a group of medical students and residents, usually shackled with high stress, tight budgets and limited free time, and then remove all three of those concerns at once. Add liquor and stir. Chaos.

But that chaos is an incredibly useful tool. Our future colleagues see who we are, not who we pretend to be, while testing our ability to know our own limits. If you fail, you're likely to become one of the stories — that person who did such and such at University of Somewhere's interview. Applicants and residents all hear stories passed through the grapevine of the intimately small world of neurosurgery. The stories are hilarious — if they happened to someone else. If you are the story, you might end up among the many who apply to neurosurgery in the fall but have no program to train with come spring.

And on day two, the haze and sleep deprivation serves another useful purpose. I'm not my best self when I'm not well-rested. Then again, I'm about to go seven straight years without being well rested. A residency program may not want to know what my best self looks like, because that's not the man they'll see much of the time. Why would they interview my Dr. Jekyll, if they'll mostly be seeing my Mr. Hyde?

Despite the hangover, I had carried the tune well enough to "pass" that morning. However, I was hardly out of the proverbial woods. Most interviews were conversational, but experience had taught me that they could be set with surprises. I had been asked to share my most embarrassing story. I had been made to tell the dirtiest joke I could think of to a group of gray-haired Gen X-ers. I had been ordered to make a sculpture with a Lego-man, a pile of office supplies, and a sixty-second time limit. I had been challenged with finding the "best video on YouTube," as well as my favorite one, as though I should have judged these categories differently. I was asked to imagine myself reincarnated at another point in history and to describe the life I would lead there. I was told to graffiti over the headshot I was required to submit with my application, assured, of course, that I would be judged on my handiwork. And I sat in that chair, guitar in hand, knowing I would have seven more interviews by that afternoon, but not knowing what each might entail.

Now, the right fit is certainly important, but there are clear downsides to this kind of feeling-out process. The constant question *are you one of us?* is tough for some applicants to answer. I don't envy the talented women who pursue neurosurgery, because the question is sometimes phrased, *are you one of the guys?* I know a gay applicant who worried about his identity's impact on his application. Another interviewee, an observant Muslim, hid behind a constantly carried cup of liquor that he would never actually drink, in order to pass the likable/sociable test. Besides the occasional Asian joke, which I'm not the type to worry about, my path was smooth by comparison.

And there are certainly healthier ways to spend a few months of your life. My typical week is beerless and regimented, fueled by Cobb salads and coffee, with near-daily exercise and a disciplined diet. Not so during interview season. The trail is a riotous road of open bars and steak dinners, where medical students, used to scrimping their student loan money, suddenly experience excesses and extremes. I flew coast-to-coast, and to-coast again — twice in the same week. For the first time, I would eat until I vomited. Then, like a Roman, I ate some more. I rediscovered why I hate strip clubs, and fell asleep standing up at least twice. I had, in rapid succession, a Chicago-style pizza in Chicago, a New York style pizza in New York, and a root canal on my second biscuspid. I had my first taste of the club scene in Iowa City — and it will probably be my last. I went to Texas, Los Angeles and our nation's capital, and was astounded by the beautiful variety of my country. And by the end, though a weight had been lifted from my shoulders, I was nine pounds heavier, and my suit had become snug.

Now, the interview trail is over, but the rest of my story is just beginning. In June, I'll have the terrifying privilege of starting my neurosurgery career at a world-class hospital in the city I call home. Pretending I got here because of my own accomplishments would mean ignoring how lucky I have been to have mentors who backed me, as well as patient support from friends and family. I don't know where I would have ended up without them, but my next seven years could have looked very different.

"That was terrible, but you pass," said Dr. O. (I know his name now). We all chuckled. And now, months later, I still chuckle to think of all that I've seen and done in the name of neurosurgery interviews. In the second half of my twenties, with degrees piling up, I'm finally a few weeks away from having a goddamn job. It is exactly the job I want, and it is one I am honored to have been offered. But while countless hours in the library certainly helped, they only got my foot in the door. I hope you never need surgery, and that your life is a healthy and happy one. But if you ever do, here's an unsettling truth to contend with — your surgeon got his first job by being a person you want to have a beer with.

The Final Year: Match Day Spotlight

Match Day Spotlight: Internal Medicine

June 5, 2017

Nina Kogekar
Icahn School of Medicine at Mount Sinai
Class of 2017

Tell us about yourself.

I'm originally from Connecticut and went to Swarthmore College for undergrad. I was a biology major and did a lot of singing in college. I went to medical school right after graduating. I've really enjoyed my experience as a student at Mount Sinai and am so happy I get to stay on as an internal medicine resident here.

Looking back on your medical school experience, what would you say to the young and naïve "first-year you"?

During first year, I considered specialties from the perspective of the diseases or organ systems on which the specialty focused and whether or not I found them interesting. However, if I could go back, I would also encourage myself to consider what the typical day-to-day activities are like for a given specialty. I would also say that first year is a great time to keep an open mind and explore a broad variety of specialties and activities.

What tips do you have for USMLE?

As a second-year, it can be frustrating when all of the advice is to "do what works for you," but it is true that everyone has a different approach. Do not be afraid to change your study methods if what you're trying isn't working for you! Doing lots of practice questions, like on UWorld, is very helpful. When doing practice questions, it is especially important to study and learn from the ones you get incorrect. Also, make sure to build in time for breaks and fun with friends during study time!

What advice do you have for the students going through clinical rotations?

Don't be afraid to speak up, even if it can feel intimidating at first. Try to find out what expectations are for medical students, as they may differ from rotation to rotation, and ask for feedback early. Take the time to talk to your patients and get to know them well. Help out and support your peers.

What recommendations do you have for medical students to maintain their sanity?

Try to keep up with your non-medical hobbies as much as possible. Make the time to hang out with friends and family, even if it means sacrificing a little study time. Spending time with non-medical school friends can help put things into better perspective.

How did medical school differ from your expectations?

I was surprised at how different the pre-clinical and clinical aspects of medical school felt. Though they complemented each other well, it almost seems like they are two different programs. My time in medical school also passed much more quickly than I expected.

What things did you do during your four years of medical school that you believe particularly impressed your residency program?

I was involved in a few different activities that aimed to provide care to underserved patients, including our student-run free clinic. On interviews, programs were interested in learning more about these experiences. I also got a lot of questions about my research project. I wrote about singing on my application and also got questions about that from interviewers; it was nice to have the opportunity talk about interests outside of medicine.

What attracted you to your chosen specialty?

I was already considering internal medicine when I entered medical school, and all my experiences confirmed that it was the specialty for me. I love the breadth of the field, in terms of the diseases, the patients encountered, and the variety of career paths available after residency. I enjoy having the opportunity to think through complex clinical problems and apply what I've learned about pathophysiology in practice. I also liked the types of interactions and relationships I had with patients on rotations in internal medicine.

What is your biggest fear about beginning residency?

The transition from being a student to being a provider with real responsibility for patients, and starting to make decisions on my own.

What advice would you give third-year students about to start the Match process?

Get advice from multiple advisors in your specialty with respect to your fourth-year schedule and programs to which you should apply. Prepare a CV to send to letter writers and give them plenty of time to write your letter. Complete a draft of your personal statement early on, so that others can look it over and you have time to make revisions. Practice answering some of the common questions that come up in interviews in order to prepare yourself for the real thing.

Match Day Spotlight: Obstetrics & Gynecology

May 25, 2017

Jameaka Hamilton
Medical University of South Carolina
Class of 2017

Tell us about yourself.

Hi! I'm Jameaka Hamilton, 25 years old, from Blythewood, South Carolina. I'm a fourth-year medical student at the Medical University of South Carolina going into obstetrics & gynecology! My parents were both in the military and were very supportive when I decided to join the Air Force to help pay for medical school. I have one younger sister who is pursuing a PhD in Entomology at NC State University.

Looking back on your medical school experience, what would you say to the young and naïve "first-year you?"

I would tell my naïve first year self to have more fun. At the beginning of medical school, it feels as though there isn't enough time to study all the material in your syllabus and have a life, so you end up compromising and miss out on special events with friends and family. Finding that balance is key so that you don't lose sight of why you wanted to go into medicine in the first place.

What tips do you have for USMLE?

Have a study plan, and stick to it! It's easy to feel overwhelmed by how much material you must review in such a short amount of time, but know that you can and will succeed doing it! Keep your support system close, and take breaks to enjoy yourself. Know that just because it's your dedicated study time doesn't mean you can't still have (some) fun!

What advice do you have for the students going through clinical rotations?

Be prepared. Be on time. Be a team player.

What recommendations do you have for medical students to maintain their sanity?

To maintain your sanity as a medical student, it is incredibly important that you maintain healthy outlets for stress! Find a hobby that you enjoy that adds value to your life. Keep in touch with your family and friends who have supported you up to this point in your career. And above all else, lean on your classmates who are going through this journey alongside you!

How did medical school differ from your expectations?

Medical school exceeded any preconceived expectations I had! It required more time management skills to manage all the material you're responsible for as a medical student while at the same time trying to have a life.

What things did you do during your four years of medical school that you believe particularly impressed your residency program?

I took the initiative to get involved in several leadership roles across campus to try to make an impact in my medical school community. I got involved with several research projects early on as I knew I was planning on entering a competitive specialty. I also made an effort to improve my mentoring skills by working with underclassmen as well as high school students interested in pursuing a career in medicine.

What attracted you to your chosen specialty?

I always knew I wanted to pursue a career in a surgical specialty even before coming to medical school, but it wasn't until my third-year clinical rotations that I found "my people." Obstetrics & gynecology offered me the balance I was looking for in a specialty during my clinical rotations — the opportunity to provide primary care while at the same time becoming as a surgeon. It also doesn't hurt that they get to help bring life into the world on a regular basis, which is pretty awesome!

What is your biggest fear about beginning residency?

My biggest fear about beginning residency would be making the transition from student to resident. Almost overnight we are tasked with much more responsibility than we've ever been given, but I'm ready for the challenge!

What advice would you give third year students about to start the Match process?

The Match process (whether with the military or as a civilian) can be crazy expensive and sometimes stressful, but it's really an opportunity to explore what's most important to you in the next stage of your medical career. Use your mentor in your specialty to help determine how competitive of an applicant you are so you can make the best decisions on the type and number of programs you apply to. Figure out what region(s) of the country you're willing to apply to, and if you're really interested in a specific program, considering doing an away rotation there to improve your chances of getting an interview offer. Go on as many interviews as you can and try to go to the resident dinners! It's the best opportunity to get an impression on how the residents like the program and get your questions answered. Take notes on what you like and dislike about all the programs you interview with to make it easier when you make your rank list.

Match Day Spotlight: Pediatrics

April 27, 2017

Allison Lyle
University of Louisville School of Medicine
Class of 2017

Tell us about yourself.

I come from a small town in southern Indiana near Louisville, Kentucky. I have wanted to be a pediatrician since I was 14. My undergrad degree was in Biochemistry and Honors Research from Indiana University. I then earned a Masters in Bioethics and Medical Humanities from the University of Louisville. After that, I got married, traveled and worked for two years in medical research at the Indiana University School of Medicine before coming back home to Louisville for medical school. My husband and I started a family in medical school, and our daughter is now almost three years old. We love hiking in the national parks and are trying to visit all 50 states.

Looking back on your medical school experience, what would you say to the young and naïve "first-year you"?

Be more confident! I applied to medical school four times before I was finally accepted, and all of those previous rejections damaged my confidence. I definitely suffered from imposter syndrome! I worried that, because of my history, I would always be struggling to keep up with my classmates. Then, in first year, I was pregnant with my daughter and I felt like everyone would perceive me as undedicated to medicine. It was a challenge, but I wouldn't have done things any differently ... other than being more confident.

What tips do you have for USMLE?

Start UWorld and First Aid early! I held off until spring semester of second year to really delve into both of them and I think it was a mistake not to start earlier, as early as first semester. What worked best for me during my dedicated study block was to stick to a schedule and not worry about how anyone else was studying. Doctors in Training (DIT) worked wonders for me, but I also used UWorld, First Aid, and Pathoma.

What advice do you have for the students going through clinical rotations?

So much advice! First and foremost, have fun! The clinical years were much more fun than the didactic years. Start studying for the shelf early, be a good team player and spend as much time as possible with your patients. And remember to smile!

What recommendations do you have for medical students to maintain their sanity?

Find your outlet and stick to it. Running, yoga, meditation, time with family — whatever it is, do it and stick to it. There *is* time. Even if it feels like there is not.

How did medical school differ from your expectations?

I knew that medical school would be challenging, but I never imagined I would be a medical student and a wife and mother all at the same time. Medical school itself is not the hardest thing I've ever done, but being a mother and being in medical school definitely is. I never thought I'd be this sleep deprived and stressed yet this *happy*. I never imagined that I would have such close friends and mentors that are so brilliant, so caring, who are just amazing human beings.

What things did you do during your four years of medical school that you believe particularly impressed your residency program?

Having interests and being passionate about something. I hold a Masters in Bioethics, so outside of my studies and rotations, I have taken an interest in perinatal ethics and communication with parents, and I let that shine in my interviews. Nearly every program asked me about it.

What attracted you to your chosen specialty?

I love kids! I'm a mom so I have lots of experience with kids already and I liked connecting with parents and being able to say that I had been there, too. Then, my very first patient on pediatrics was a one-year-old who wanted me to hold her and play, and I was hooked. Kids are the best!

What is your biggest fear about beginning residency?

That I don't know everything and will hurt someone — but that's why there is supervision!

What advice would you give third year students about to start the Match process?

Use your resources! My home program's program director was amazing about helping me in the selection process with how and where to apply. My advisor was also a wealth of knowledge and expertise in helping me with deciding if/when/how to cancel interviews. Start writing your personal statement early and have lots of eyes read over it. Go on a mock interview and prepare with standard interview questions so that the words don't sound weird coming out of your mouth the first time — you don't want to sound rehearsed, but you do want to sound confident in your answers. Pick one or two things that you want the program to know about you before you leave and find a way to work that into your answers to interview questions. Go with your gut when it comes to ranking.

Match Day Spotlight: Psychiatry

May 28, 2015

Laura Black
University of Washington School of Medicine
Class of 2015

Tell us about yourself.

I grew up in the Seattle area and studied Spanish language and literature at Western Washington University. Before medical school, I spent two years living in Spain studying Spanish and taking medical coursework. I then returned to Seattle and got involved with a local bilingual community health center, performing health disparities research and volunteering as an interpreter. I came into medical school at the University of Washington with a strong interest in health disparities, especially those affecting immigrants and refugees.

What residency program will you be joining and where?

I matched at NYU's psychiatry program in New York City and will be training at Bellevue Hospital.

Looking back on your medical school experience, what would you say to the young and naïve "first-year you"?

You might feel like the most ignorant person in the room (I sure did!), but you bring a perspective and experience to the table. Part of your job is absorbing the knowledge imparted to you by seniors, but another part is bringing a new set of eyes and questions to the situations you encounter. In basically any situation (educational or clinical), it's probably best to keep your questions open-ended and be careful of assumptions.

What things did you do that you believe were valuable to succeed the first two years in the classroom?

Finding a study method for medical school can be very stressful, and I think students sometimes feel pressured to reinvent the wheel. Remember that you know yourself best; try to draw on study methods that have worked for you in the past. For me, that meant lots of verbal review with study partners, as well as kinetic learning by drawing and writing on a white board. But one method seemed to hold true for everyone — do a lot of practice questions!

What things did you do that you believe were valuable to succeed the second two years through clinical rotations?

Try to identify a goal for each rotation, like gaining confidence in the neuro exam, giving more organized oral presentations, and so on. Find something to be excited about, even if it's just the fact that you may never be in the OR or deliver babies again. Residents and attendings are not expecting perfection! But they do appreciate you being punctual, professional, and asking here and there for tips on improving your presentations. Pay attention to residents' presentations for things to emulate. Help with discharge planning as much as you can — you'll often have the most time to think about your patient's social situation. Ask your senior resident for help with particular physical exam findings or maneuvers if you're not sure or just want extra practice.

What things did you do during your four years of medical school that stuck out or particularly impressed your residency program?

Pursue your passions. I continued working with Spanish-speaking patients because that was my interest, and I did a summer internship in rural family medicine with migrant farmworkers. I also spent time at a hospital in Paris working with refugees. These were both great learning experiences, and they communicated to residency programs that I wanted to continue on the path of community health and global health disparities.

What things were unhelpful or you wish you hadn't done in medical school?

Don't think you need to be involved in everything! You need to sleep and perform well in your coursework. Get involved in a select few things you care about. It's perfectly okay for most of your major activities to be during non-class times, like during the summer or time off of rotations. You don't need to do everything simultaneously, though some of your classmates will do that, which is fine, too. Remember that a huge component of your residency application will be clerkship grades and evaluations, and that clerkships are also where you will likely decide which specialty to pursue. So, during the preclinical years, do get involved in areas you think you might have an interest in, but don't feel pressured to have a prepackaged resume by the end of second year. Be open to being surprised.

What was your level of involvement in research and other extracurricular activities, and your opinion on how important that involvement is?

I think it's easiest to get involved if you do a formal research program, usually the summer between first and second year. I think this can be valuable and is a good way to meet mentors. I didn't become involved in research until later, when clinical mentors got me involved in some ongoing projects. I think the importance of this depends on your specialty.

What attracted you to your chosen specialty?

I was interested in community health and initially was headed toward primary care. However, I was most drawn to patients with mental health and psychosocial challenges. I found that my primary care mentors often wished they had more time and training to work with these patients. I spent some time with a child and adolescent psychiatrist at a residential treatment center and was immediately drawn to both the patients and the psychiatric approach.

What attracted you to your residency program?

I hoped to find a program with rigorous academics, breadth of clinical training, and a strong sense of social conscience. NYU was a natural choice. Bellevue is the oldest public hospital in the United States, and I really connected with its spirit of serving patients from all walks of life. New York is also a great place to learn psychiatry with many conferences, professional organizations, and psychoanalytic institutes.

What things did you do to maintain your sanity in medical school?

Lots of yoga and pilates whenever possible! My classmates and I would also get together just to make silly crafts or do yoga. Even five minutes of a meditation podcast, or a 10 minute jog. Little things add up, and that can be hard to remember when you're constantly pushing yourself to overachieve.

The floor is yours — what do you wish to share with current medical students?

Take it one step at a time, and trust the process. You will do great.

Students recognize that medical school is transformative when they see how far they've come — hearing more students stumble over early presentations, or no longer having to rely on mnemonics to make decisions. Truth be told, however, there is growth at every step of the way. Amara Finch begins medical school wondering if she's in the right field — a surprisingly subversive question among medical students — and over the course of the year, returns elegantly to her feelings about uncertainty. Though this doubt is stigmatized in some classrooms, I would venture to guess that most students reflect on this idea more openly with their friends and family outside of school. In "On Being a Medical Student" and "The Measure of a Medical Education," Brent Schipke and Ajay Koti reconsider ideas about personal growth — for example, what is and isn't promised to students as they embark towards their degrees? It can be a shock when knowledge and training cause you to lose a part of yourself. Finally, Thomas Yang Sun and Sarab Sodhi, about to graduate, take a few moments to look back on what the process was like.

The Journey

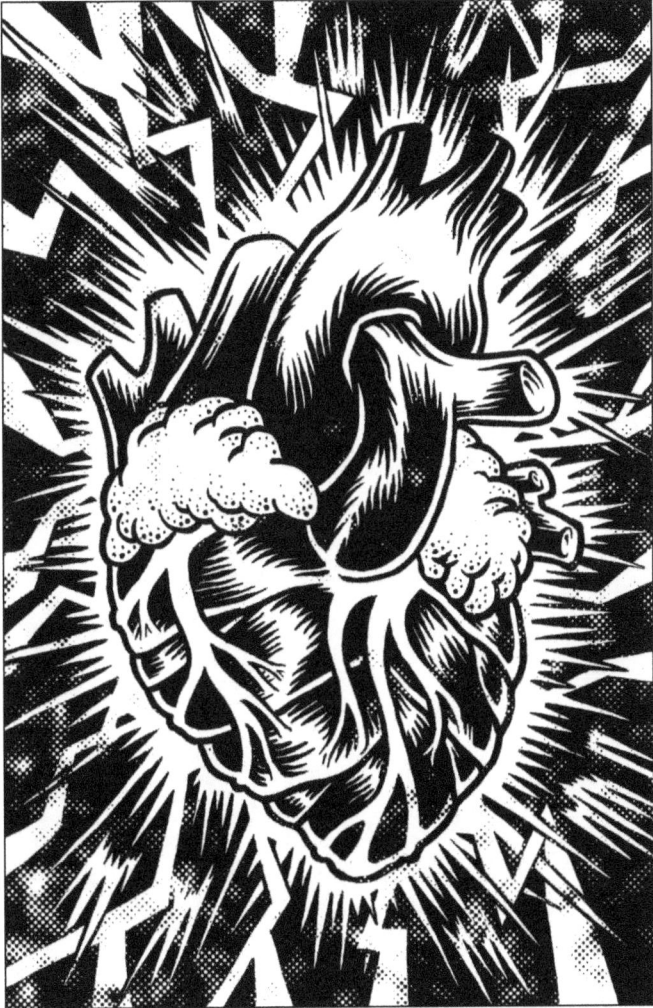

Electric Heart by Billy Sveen
Class of 2015 at Loyola University Chicago Stritch School of Medicine

Transformation from md to MD

June 8, 2017

Thomas Yang Sun
Yale School of Medicine
Class of 2017

First year of medical school:
Don't remember much.
MD/PhD students, you know what I mean.
Learned how to use a stethoscope.

Second year of medical school:
Everything a blur except
Step 1 introduced me to my friends
melatonin, Benadryl, Ambien
And my best friend
Lunesta.

Third year of medical school:
First clerkship: Ambulatory.
First time I saw a patient by myself!
It took an hour and a half!
Attending happy?
Attending not happy.

Second clerkship: Neurology.
Oh Neuro.
So hard to "localize" my feelings about neuro.
Where is the lesion, Thomas? My attending would ask.
Because that's very important!
Right...

Third clerkship: Pediatrics.
First time I made a patient cry.
Also the first time a patient made me cry
Actually no, just kidding.

Fourth clerkship: Internal medicine.
I matched into IM.
Enough said.

I am going to skip OB/GYN and Psych.

Seventh clerkship: Surgery
A surprisingly enjoyable clerkship
But maybe that's because I did ophthalmology.
Just kidding. Surgery was fantastic.
Until the day I stepped on the suction cord.

Emergency medicine:
Where I learned chest pain = troponin, EKG, CXR, D-dimer, CBC, CMP,
U/A, LE Doppler, abdominal U/S, troponin
Oh, and of course CTPE

Anesthesia:
For some reason I was always paired with the same resident.
She only had two weeks left before she finished residency,
So of course her favorite thing to say to me, every morning, was:
"Oh, Thomas, I can't. I can't. I can't."
And then I tell her,
"You can you can you can."
But what I was really thinking was,
"We can't, we can't, we can't."

Time out from medical school, research year:
Wow! Look at this graph! It's looking good!
P value? 0.06.

F**k!
All of my mice died.

I can't, I can't, I can't.

Longitudinal primary care clinic:
First patient of the night!
Just going over her problem list on EPIC, I saw the following:
Asthma, HTN, HLD, morbid obesity (saw that coming), pain in LLE, GERD, depression (easy, SSRI), Diabetes type 1.5 (what?), rheumatic mitral stenosis (uh oh)
And that was only the first patient of the night! And her EPIC problem list.
Plus, she checked in at 4 p.m. for a 6 p.m. appointment.
And she's scheduled to see me for a good, solid 15 minutes.

O.M.G.

Fourth and last year of medical school:
Beginning of my medicine sub-internship:
On my IM sub-I. I got a page! From the care coordinator, whew.

I got another page, from a nurse, saying "come to pt room N.O.W." I grabbed my stethoscope and ran to the patient, whom I was cross-covering for the night, and whom I did not know and had never met. Nurse was now telling me, patient was blue in the face, and desatted to the 70s.

O.M.G.

The respiratory therapist was also in the room, and as soon as she saw me, she said "Patient is okay now, I gave her NTS."

Now, I didn't know what NTS was, but I knew it saved the patient. So I didn't ask any questions.
I went up to the patient, and pretended to listen to her lungs even though I had no idea what I was doing or what had just happened.

Fast forward to the end of my sub-I:
I got a page. It was from a nurse. "Ms. M is not looking well, she's very somnolent," it says. I ran to my patient Ms. M. She had severe CHF and COPD and she was barely responding to me.

O.M.G.

But this time I knew what to do. I called my senior.

I suggested that we get an ABG immediately to check for CO_2 retention which I suspected was causing her somnolence.
(Okay, I suggested a VBG and my senior said no and told me to get an ABG, but that's not the point).

The point is, the ABG confirmed severe CO_2 retention, and we treated that by transferring her to the MICU.

Towards the end of my sub-I, on that day, I thought I could do this.
I could be a good doctor.

Soon I will be sitting at Commencement,
Having traveled the long road from a mild dork
To a Medical Doctor.

To all of my amazing classmates and
The incredible, persevering medical students
About to graduate from all over the country,
Who simply refuse to not take another exam

I am going to say to you all,

We can, we can, we can.

Differentials
August 19, 2017

Amara Finch
Emory School of Medicine
Class of 2020

Summer

It didn't take long for the truth to come out. We had just completed our second week of medical school, the anticipated "Week on the Wards" in which each student matches with an attending physician in an experience that officially marked our transition from layperson to health care professional. "From now on," our deans told us at orientation, "society will see you as a doctor. Sometimes you may not feel like one, but that is what you are becoming. This week marks the beginning of that transition, which will continue in the months and years to come."

That Friday afternoon, I sat with a group of women at a table outside the medical school to debrief. The air was thick with the heat and humidity of a southern summer, and though we hadn't known each other for long, the jarring nature of a new experience and numerous hours spent sitting through orientation lectures lent us the illusion of intimacy. We compared notes from our assignments: One person had observed surgeries at the VA, another had been on internal medicine at the city's public hospital. I had happily followed the infectious disease team at the esteemed children's hospital. We chatted about the long hours and our attempts to explain to residents and nurses that yes, we are technically medical students, but no ... please, please don't expect us to know anything yet. We recounted highlights, interesting cases and the intimidating thrill of the fact that we had really made it to medical school.

I don't remember who said it first: "I'm not sure I want to do this anymore." The bold honesty rang out among the hum of positivity. We hadn't yet started classes, and the uncertainty was beginning to surface. "I don't know," she continued. "It was cool to be with the doctors, but ... I'm just not all in. I almost wonder if I should get out now before it's too late." A few others nodded in agreement, laid out their own misgivings like offerings. I had spent each lunch break that week on the phone with my then-boyfriend, wondering aloud if I'd made a grave mistake. "What if I don't want to do this anymore?" I'd asked him. "I should have done nursing," I said one day, and the next day, "maybe I should have been a teacher." Now, my classmates and I remembered our past jobs: consulting, farming, research, advocacy, education. The positions we'd worked hard to move away from suddenly seemed alluring, and we wondered why we'd left. One week earlier we had learned the term "escape fantasy" from a faculty physician, and we discussed our own premature escape fantasies, feigning humor to disguise anxiety. We imagined alternate futures, populated by versions of ourselves that seemed, already, to be slipping out of reach.

Autumn

As the seasons turned, we began to learn the language of medicine. We studied words for things we'd never known to name, to appreciate nuance in what had once seemed straightforward. *Simple squamous, cuboidal, transitional, columnar.* I'd never known our casings to be so complex. We began to match symptoms to processes and processes to pathologies until we could, by autumn's end, try our hand at diagnosis. Our small group sat with printed copies of chief complaints, lists of vital signs and associated symptoms. We learned to craft a differential diagnosis, a list of diseases that could cause the suffering described in these fictional patients. It felt like a game: We would try to find a possibility in every category, imagine the many biological stories that could be occurring inside bodies we would never see. Our untrained minds ran wild with ideas.

Though the art of crafting a list of differentials was new, the concept felt familiar. I had been turning over a list of my own possible paths in the back of my mind since the summer. The categories I populated were not based on organ systems; instead, they represented people, places and professions. I nurtured some steady fantasies: working as a teacher, a social worker, a nurse. These were the horses, my most likely differentials. Occasionally I would entertain zebras, imagining my life unfolding as an ecologist, a journalist, a park ranger, a cook.

As the days grew darker, we learned the art of dissection. I found myself in the medical school more than in my apartment and felt less at home in each. The longer I lived in Georgia, the more I cultivated visions of living in Colorado or California or Alaska, the late nights in anatomy lab tempered by distant ridgelines and the curves of coasts I've never seen. I envisioned alternate endings to my relationships, imagined staying with the love I'd left behind. I fantasized about leaving school, starting over. I practiced the lines I would use when explaining it to my family and friends. At first my mind's wanderings were confined to the quiet of the evenings as I studied and lay down to sleep. But the meanderings metastasized until they clouded my mind during lectures, meetings, afternoons in clinic. The days became blurred by thoughts of a thousand futures unrealized, futures that had or had never lured me before.

Winter

I know now that I was never alone. Initially, the students who surrounded me seemed so sure of their decision to come here, to walk into those wards and begin becoming doctors. Perhaps some of them were. But for every person who seemed so convinced of their decision, another wore the weight of uncertainty, embarrassed, as I was, by their confusion. These truths came out slowly, on long days when we arrived at school before sunrise and left long after the cool night returned. We asked in low tones, "How are you holding up?", aware of each other's fragility. And always there was an air of guilt: We would explain our unhappiness while expressing our gratitude, acknowledging the paradox of displeasure and privilege. We sent each other photographs of signs reading "Now Hiring!" at coffee shops, clothing stores, and car washes, with messages that both teased and tempted. When my family asked how school was going, I simply told them I was working hard, though I never felt I was working hard enough. Well-meaning friends outside of medicine told me they couldn't fathom committing to such a lengthy and arduous path. I laughed, silently thinking, *"Neither can I."*

Many of us were drawn to medicine for its promise of flexibility, of open doors. As the year came to an end, we could almost hear the soft clicks of doors closing behind us. In a particularly low moment, a group of students admitted that they had wished for accidents that would render them unable to continue medical school but free them from the shame of consciously quitting.

It is a common question in medical school interviews: What would you do if you could not be a doctor? I had been warned before I heard it. "Tell them you'll try again," I was instructed. "Say you won't give up." But what had seemed like a threat during interviews now sounded like an invitation. What *would* I do? Many people who had struggled to answer that question a year ago now had countless replies, could name other plans and different paths that were often vague, but always somehow better.

Spring

The warmer weather brought a welcome change: the first stages of understanding. The words we had practiced pronouncing in September now rolled effortlessly off our tongues; the heart became connected to the lungs and the liver and the skin in a system that was still opaque, but increasingly elucidated. I no longer felt like an outsider in the clinic. Not quite a citizen, perhaps, but also not a tourist.

To study medicine is to study trade-offs, to become intimately aware of the costs of success. We learned early on that there are no perfect remedies; even our best medications have the potential to do harm. As we study pathology, it is clear that the crooked path of evolution has left us with bodies that thrive or fail depending on context. Our tendency towards salt retention has allowed us to survive in environments with limited resources; now it manifests as hypertension. Our appetite for sugar, so crucial in times past, leads to dangerous diabetes. And so it is with the most human of traits: imagination. Our capacity to envision alternate futures carries us through doubt and allows us to plan for the future. Surely this, along with our physiological adaptations, has sustained us through the harsh realities of many millennia. When uncontrolled, however, it can cause more suffering than salvation. When our minds become restless and wander, lonely among ideas of various futures and fates, we must tend to them like the other cravings born of humanity's past. We must honor and understand our imagination's presence, and learn to discern true hunger from ancestral appetites thrown out of context in a modern world.

The journey ahead will feed our imaginations day in and day out. We must nourish our imaginations; they are what will allow us to see patients in new lights, to initiate creative therapies and to think beyond the confines of medicine's established framework. On the hardest days, it will help us remember that things will get better. And it will, undoubtedly, cause us turmoil. I see this in older students, residents, fellows, attendings. I hear murmurings of "What if?", some simply musings on lives unlived, some despairing statements of resentment or regret. I've seen simple musings become realities, when people choose to walk a different path away from medicine and are all the happier for it.

We live in a culture that celebrates certainty and encourages decision. We play into a myth that we know what we're doing and that the harder we had to work for it, the more certain we must be. In reality, it is never so simple. Many of us are molded by misgivings: on good days, they are fueled by curiosity; on bad days, they are the sequelae of discontent. We live with the consequences of the choices we made, as well as the ghosts of the ones we did not. I sometimes imagine a world in which our résumés listed not only what we have done, but also the opportunities we considered but did not pursue and the experiences we gave up in exchange for others. Are those experiences, pondered and passed by, not as intrinsically important to the people we have become?

A friend once told me that the extent of our serenity is measured by our ability to let go of the things that we are not meant to have. It is a difficult task, to let go with grace. And while we find occasion to practice letting others go, we seldom think of applying the same mercy to ourselves, of releasing our grips on alternate versions of our own lives. It is an exercise in self-acceptance, in trusting the paths and process we each took, strategically or serendipitously, to arrive here.

Onwards

A list of differentials represents a beginning. It is the physician's task to chip away at the list, to carefully consider each idea in the context of what we know and do not know and cannot know. We may order tests or treat empirically. Sometimes we will rule options in and out until we achieve a diagnosis; other times we may find a cure without ever fully knowing the culprit. In some instances, we might arrive at a conclusion only to look back, weeks or months later, and realize we had been wrong. We may start over, re-think, alter course.

And so it is with us. We are dynamic. We can rule in and out, employ processes of inclusion and elimination, change our minds and ideas and plans until, eventually, we find enough comfort and confidence in our decision to keep on moving forward.

Becoming a Doctor

May 3, 2015

Sarab Sodhi
Temple University School of Medicine
Class of 2015

"WRITE YOUR NAME ON the paper," he said. Since he was a senior who'd just gotten into medical school, and I was a simple sophomore who'd chosen to attend the session, I did. "Now write Dr. in front of it." I complied. "If you're reading that and you don't feel anything, medicine isn't for you," he said. I looked at it again, my name with a Dr. in front of it. I didn't feel a thing. I crumpled up the paper, chucked it in the trash and didn't give it another thought.

Until today, that is.

In four days, I'll get to write a Dr. in front of my name. More than I get to call myself a doctor is the fact that I get to be one. In a few weeks, when I start residency, I'll be responsible for people's lives. And that is terrifying.

It's been a long journey since my careless sophomore days. I went from being a cocky, know-it-all college student, patted on the head for my intellectual acumen, to a terrified, foppish first-year medical student who spent my first year lost and confused. I was petrified in anatomy, as I was constantly less aware than my classmates who with reckless abandon pointed out the vagus nerve, the mesenteric arteries and dismissed much of the fascia I believed was important anatomy. I drowned in the weight of neurophysiology, as I discovered that the brain was and remains a complete mystery to me. I threw up in the hotel before I went in to take my Step 1 exam and was mortified when I barely made an average score. I spent my first years of medical school battling the terror of inadequacy, afraid I wasn't good enough or capable enough.

But, in four days, I'll be a doctor.

I was terrified that I was playing doctor this entire time. Mortified that my medical school experience was not enough, that I was unprepared for the next step. Then, I realized something.

I've diagnosed and initiated the management for dozens of diseases. I've read hundreds of EKGs and chest x-rays. I've brought life into this world with my own hands, and been there when it's left. I've fought violently against death, breaking ribs as I tried to bring back a patient from the precipice. I've watched death softly take someone who was ready to go. I've cried for a patient in the arms of my lover after I first told someone they were going to die. I've violated the sanctum of the body with chest tubes and central lines in hope that someone would live.

My family, like families tend to, have introduced me as a doctor for a few months now. I've demurred, each time saying, "I'm not a doctor yet." Like my white coat ceremony, I need something to mark the movement from a medical student to a physician, and to mark the importance of the situation. I realize, however, I'm not there yet. Though I'm getting my degree, I have, in Robert Frost's words, "miles to go before I sleep." And I'll never be there. I'll constantly be learning, making mistakes and fixing them, and forever humbled by the vast enterprise of medicine I've had the audacity to try to conquer.

I've been becoming a doctor for a long time now. The MD I get to put at the end of my name has been in the works for eight long, caffeine-fueled, sleep-deprived years. I'm going to spend the rest of my life living up to the promise it holds, though — because that's what becoming a doctor really means.

The Measure of a Medical Education: An Epilogue

June 26, 2017

Ajay Koti
Morsani College of Medicine at the University of South Florida
Class of 2017

"WHAT'S THE MATTER WITH everybody?" asked Mrs. Palmer, a hopelessly demented woman with water wells for eyes. She had just endured her third consecutive tongue-lashing by the bulldog masquerading as a nurse anesthetist. Her bewilderment only slightly exceeded my own; in a surgical suite on my first day of clinical rotations, I was mentally rehearsing how to scrub in without accumulating any (more) ire from the operating room staff.

That all seems so long ago now, sitting in a hotel lounge on my final stop of the residency interview trail. I had completely forgotten Mrs. Palmer until I began flipping backward through my journal to criticize my old writing and lament the downfall of my penmanship — from fluid scrawl to deranged hieroglyphics. I skimmed the indecipherable pages, succumbing to waves of nostalgia; once the nostalgia subsided, I realized that the notebook was a comprehensive record of my medical education. This raised a new question.

How could I measure that education?

Maybe my medical education can be measured in student loan debt, the size of which long ago passed the limits of my comprehension. Perhaps I could measure it by my exponentially increasing collection of gray hair (though still a minority, if the pace keeps up, I will go full Anderson Cooper by my mid-thirties).

At a minimum, the last four years should reflect the accumulation of medical confidence. I spent precisely one day celebrating my acceptance to medical school; every day since has been consumed by some degree of "impostor syndrome." Surely the admissions committee had made some critical error — it was only a matter of time before medical school would expose me for substandard intelligence, misplaced ideals or any other of a litany of deficiencies. But maybe if I wore that short white coat and acted the part, I could convince people that I belonged. I might even convince myself.

But the coat, and whatever authority came with it, only sustained my insecurity. Patients divulged deep, personal secrets to it. Strangers acknowledged it with a smile and a "Hey, Doc!" in hospital corridors. One even paid for my lunch when I found myself wallet-less in a cafeteria line, dismissing my objections with, "You take care of others; let me take care of you." Each was an undeserved courtesy, and I wanted nothing more than to yell back in a blaze of melodrama — I'm a fraud! My only bulwark against being debilitated by impostor syndrome were two short words of solidarity from classmates: "Me too."

I returned to the notebook for answers. It held essays on pre-clinical lectures I barely remembered, short write-ups of bad first dates, and retrospectively painful jokes about Donald Trump. I found a bad poem about the cadaver lab, and a worse one written as an ode to ballpoint pens stolen by attendings. There were scores of patient vignettes and dozens of half-baked soapbox manifestos on medicine and humanism. Slowly, more than the handwriting began to give me trouble. I found the writing well-intentioned but often simplistic, naïve. At times, I barely recognized myself. The journal was a record of how I had changed over four years.

And there was the answer — it was my own transformation that measured my medical education. I'm more thoughtful now and more prepared to traffic in nuance and compromise. I'm more confident in front of patients and non-medical friends, and less confident among peers. I'm more focused and more practical. I'm also a little crankier, more impatient and more reclusive. I can, and have been, prickly and arrogant to family members, particularly when they engage me on medical issues.

I don't know how to assess these changes — good, bad and ugly — or their aggregate. I'm still too close to the experience to make any credible judgments.

But perhaps such a judgment is unwise anyway. For each scribbled-in page of my journal, there are untold blanks waiting to be filled in with more of my own transformation in the coming years. My medical education remains incomplete, and as I proceed through graduation, residency, fellowship and whatever comes thereafter, I will continue to change, in ways that I cannot yet appreciate. Hopefully, that change will be for the better.

The Making of an American Doctor

November 6, 2016

Matthew Trifan
Lewis Katz School of Medicine
Class of 2017

NOT LONG AGO, I was on duty in the emergency department, sewing up a kid's lacerated hand. He was ten years old and terrified. I had to make all kinds of promises to numb him up before starting. As I cajoled him, I had the strangest sense of déjà vu. I realized that I had lived through the same experience myself — as a young boy sitting in my kitchen with a torn-up hand, having careened on roller-skates into a pile of rocks. Only the doctor had been my father, and he had coaxed and pleaded with me just like I was doing now. I remember the burn of the lidocaine, and then being mesmerized by my father's deft weaving of knots. Now here I was: spinning the same knots, singing the same song, soon to be a doctor myself. Just like that, our lives had swung full circle. The torch had passed on.

I find myself thinking about my family a lot these days. And for good reason. My journey to becoming a physician really started with their own.

—

My parents' story begins some fifty years ago in Romania, a beautiful country nestled in Eastern Europe's Carpathian Mountains. In the 1950s, Romania was yet another country yoked to the Soviet empire, bearing the Communist flag. My mother was born a city girl in the capital of Bucharest. My father was raised in the smaller Transylvanian town of Brasov. As teenagers in the early seventies, they grew up with a love for all things American. They jammed out to black-market records in their rooms, from CCR, to Pink Floyd, to Jimi Hendrix and Fleetwood Mac. They wore "blue jeans" out in public. They indulged in the rare Pepsi-Vodka at the discotheque. Everything American was edgy, rich and cool, and, more importantly, subversive to the Communists stiffs who ran the country.

What little they actually knew about American life came from books and movies smuggled into Romania. My parents learned about Western culture from films on the black market. These movies were bootlegged ad nauseum and watched with neighbors and friends in living rooms at night. My mother loved the forbidden love saga of Natalie Wood and Warren Beatty in "Splendor in the Grass." My father was particular to gangster movies and hardboiled crime noirs like "The French Connection." The heroes did what they wanted and said what they wanted, wherever and whenever they wanted. "We couldn't believe how bold they were," my mother recounts. "We couldn't get enough of it."

It wasn't until the end of the 70s, when the deprivations of the Cold War began strangling the Romanian people, that the "American dream" became an escape plan. Suddenly there were food shortages, crackdowns on political dissent and, most obnoxiously, long waiting lines for everything! My mother would wait three hours in line to buy butter; my father did the same for toilet paper. It was no longer easy to laugh off "incompetent" government bureaucrats — not when you couldn't put food on the table, or get petrol for your car or find a steady job. More and more Romanians were spying on each other, acting as informants at home, in school, in churches and even in hospitals. Daily interactions became hostile, suspicious and dangerous. Some citizens tried to emigrate westward, but even this became difficult. Obtaining travel papers was an increasingly onerous process, taking months to years to complete. If someone was lucky enough to travel abroad, their family members might be forced to remain in Romania — to reduce the risk of defection. The country was turning into a prison.

One Romanian finally had quite enough of this. In the late 1950s, my grandfather, a surgeon, had made a careless toast at a dinner party — saluting the "Hungarian Revolution" against the Communists — and an informant had reported him to the police. Thus, my grandfather received a ten-year sentence as a political dissenter in a laboring gulag. This was the type of sentence that often meant death from malnourishment, guard brutality and disease. Luckily, his medical knowledge proved to be an asset in prison. Because the nearest hospital was six hours away, the guards came to depend on him for his medical expertise.

My grandfather had been prisoner for about five years when the prison warden approached him one night in a state of panic. The warden's wife experienced a tubal pregnancy, which had ruptured. She was hemorrhaging blood and would not survive the journey to the hospital. "Do everything you can to save her life," the warden begged him. Without hesitating, my grandfather pillaged the jail's shoe-repair shop for cutting instruments and thread. He boiled the instruments to sterilize them. Then he operated on the warden's wife. He stabilized her bleeding, and she was able to reach the hospital. In gratitude for saving her life, the prison warden commuted the remaining sentence. My grandfather emerged from the gulag in 1963 with a clean slate to practice medicine freely. He resumed his career as a surgeon.

But he would never forgive the State for robbing him of his children's earliest years. My father was seven years old when my grandfather finally came home.

Thus, in 1978 — fed up with being spied on, underpaid and overworked — my grandfather obtained travel papers to America. He was the first of his family to leave Romania, albeit under the guise of returning soon. Upon arriving at a friend's home in Ohio, however, he applied for political asylum. The US government reviewed his history and granted his request. There was a steep price for admission. Because of his poor English and lack of an American degree, he had to surrender his profession as a surgeon. He would finish his career as an OR technician in Ohio, relegated to fetching instruments and assisting surgeons from the sidelines. He took comfort in his medical expertise, which won the respect of his colleagues, and more importantly, in his job, which meant he could host his family in America. Using a Romanian emigration law known as "family reunion," his wife and daughter were able to join him in the United States.

Now there was only the issue of my father. In 1981, at age 25, my father was the last of his family left in Romania. He had his heart set on America, but leaving would not be easy for him. For starters, he was madly in love with a woman, who had her own career — and her own family — in Romania. He met my mother during his mandatory service in the Romanian army after high school. She was the sister of his best friend in the army. One rainy Sunday afternoon, she had come to visit her sibling at the camp, and my father had trudged along to say hello. He was thunderstruck. At the time, she was dating another man, but that did not deter my father. He was dogged in his courtship, catching weekend trains to visit her and writing love novellas to mail her, dozens of pages at a time. Every night, after marches and drilling, he dreamed up a life of them together on ink-blotted pages.

By 1981, my parents were happily married and miserably broke. My father had finished medical school in Bucharest. Despite his budding career, he had his eyes set on America. During the next two years, the Romanian economy continued to crash, and my parents decided it was time to leave. They began lobbying government officials for travel papers. Strings were pulled. Favors were called in. It was a long and expensive process, lasting a full year. Eventually they got their wish in the form of a travel passport via Italy. There was no time to waste. My parents packed lightly and gathered what little money they could. They were permitted two suitcases each and $50 per person for travel expenses.

Their journey was harrowing. My father was convinced they would be stopped by the secret police at the airport. "When your mother and I were walking to that plane, I expected a hand on my shoulder," he said. "I was waiting to hear a friendly voice saying, Tovarăş (Comrade), where are you going so quickly?" Their sense of dread followed them into the plane, where they sat sweating in their seats. Ominous clouds darkened the skies. The plane sped down the runway and ascended in the midst of a terrible storm.

"It was as if God himself was denying us passage out of hell," my father said. Ferocious wind ripped at the plane. They held each other's hands.

The dingy aircraft rattled its way to Albania, where it stopped for refueling. The passengers waited inside a tiny airport during the storm. Through a huge window, they could see a dozen giant tanks parked along the darkened runway. The mammoth turrets were pointed directly at them. They waited for hours in the airport with their hearts thundering.

The plane was refueled, and a few hours later my parents arrived in Rome. They bought Italian gelato from the nearest vendor and sat down on the sidewalk. They held hands. With the warm sun beating on their necks, they wept.

—

It would take nearly a month for the American embassy in Rome to grant them entry visas. In the interim, my parents, then ages 28 and 30, received $200 from my grandparents in America. The money allowed them to travel to the island of Capri off the gulf of Naples, and then to Florence. This was their impromptu honeymoon of sorts — their first true taste of love in the free world.

In the spring of 1984, my parents landed at JFK International Airport in New York City. They had made it to the Land of the Free — only to find a fresh set of hardships. They were broke, with less than 100 dollars between them. Although my mother was fluent in English, my father spoke miserably little of it. My mother began work as a bank teller for Citibank for $800 per month. My father discovered that his eight years of medical training in Bucharest held no weight whatsoever in America. He would need to apply to residencies without a formal US medical degree — and with minimal English to boot! His applications were rejected by nearly 30 residencies in the region. On the phone, he was bluntly informed "We do not take foreign graduates." He was dejected. No longer the breadwinner, he sat alone in the apartment most hours of the day, confronting the choices he had made.

He caught a break through a fellow Romanian in New York. My father was offered a "voluntary internship" at a New York hospital for a year, unsalaried, as a "trial" for residency. His English was rudimentary, meaning he would have to learn the language as he went along. And he did. He impressed his fellow residents and physicians with his acumen and astute physical exams. He could diagnose pneumonia almost exclusively from pulmonary percussion — at a time when his colleagues were mostly relying on radiographic imaging. He likewise dumbfounded an infectious disease specialist by diagnosing tuberculosis by stethoscope alone. Half a year later, my father was offered a residency position at the hospital. He had made it in America.

With a residency secured, my father and mother felt it was time to begin a family. My older brother was born in 1987. I was born in 1989, the year the Berlin Wall fell. It was also the year that the Communist regimes of Eastern

Europe began toppling like dominoes. On December 25, 1989, the former Communist dictator of Romania, Nicolae Ceaușescu, and his wife Elena, were dragged before a kangaroo court and charged with genocide. My parents watched dumbfounded as the proceedings unfolded on television. The Ceaușescus were found guilty and sentenced to death by firing squad. They were taken behind the courthouse and lined against a wall — like so many of their victims before them — and shot to death. "Our whole lives they were untouchable," my mother recounts. "Then, poof, just like that, they were gone."

A new era was dawning in Romania — and in my parents' lives too. In the early 1990s, our family moved to central Pennsylvania, into a house with a spacious yard in a quiet neighborhood. My father began working for the Veterans Hospital in Altoona, proudly serving veterans for the next 25 years. My mother devoted herself diligently to raising my brother and me as typical American kids. She drove us to school meetings and tennis matches and karate classes. She helped with sleepovers and birthday parties and prom suit shopping. Most importantly, she taught us the value of a dollar hard-earned, which she truly understood from her life.

Today, my brother is working happily as a hospitalist physician in Pittsburgh. My father is approaching retirement and looking forward to the most American of pastimes: golfing. My mother still travels yearly to Romania to reunite with old friends. Before passing away ten years ago, my grandfather was also able to return to the country of his birth — this time, as a free American citizen.

Now I am nearing the end of my own eight-year journey through college and medical school. Residency looms ahead. It's a fresh start in a new city. I find myself standing on the precipice of an unknown future, just like my parents and grandparents before me. Their journey gives me strength. If they could start over, I know that I can, too.

So here I sit, sewing up a young boy's hand, feeling the pinch and the pull in my own, hearing my father's voice in my head, echoing the past. My patient has calmed down. He watches me with trustful eyes. I draw each circle tightly closed, lending the wound a little more strength, before my needle travels on. The path forward is uncharted — but somehow, I know I'll find my own way.

On Being a Medical Student

September 18, 2016

Brent Schnipke
Boonshoft School of Medicine at Wright State University
Class of 2018

E ARLIER IN THE SUMMER, I was speaking with a friend from medical school while we were studying for Step 1, the big test taken by medical students at the end of second year, and he remarked, "There's really nothing quite like this. We probably don't even realize how strange it is since we're so ingrained in it." He was right: The demands of medical school often make it an all-encompassing undertaking, one that can be difficult to explain to those outside it. We try to explain anyway because it's all we have. That's why so many of us have chosen to contribute our stories to the *in-Training* community and others. What follows is my attempt at one such reflection.

Being a medical student is complex because of many conflicting tensions. One example of this tension relates to the label of "medical student." We are never far from the temptation to let this label define us. Much of this can be attributed to the nature of medical education: immersive, time-consuming and, at times, academically and emotionally difficult. Consequently, our friends, family and significant others come to accept a norm that we are always busy. This can be complicated by the fact that many of us often *like* the way that this defines us. We sometimes like being the crazy people who stay up all night to study, joking about our coffee addictions and bragging about how little sleep we get.

However, the tension arises because we just as often *don't* like what we are doing. We often *are* too busy and it is sometimes really difficult because we can't attend social events and family gatherings. We really do study, quite a bit. Each new challenge is a reminder to be humble; there will always be more for us to learn. We may not always enjoy studying, running to the hospital at strange hours and more, but we justify it by reminding ourselves that we are in one of the most prestigious degree programs in the world, and so we take pride in our profession.

However, I want to get at a more significant tension. Almost daily, we live in between "not yet" and "here we are, right now." The tendency for most people — including doctors themselves — is to view medical school as a stepping-stone, a rite of passage, an in-between time. I feel this myself; I often think of what I am doing only as the necessary steps to reach a goal. It is often just the means to an end. Speaking with residents and doctors, I find the belief is even more pronounced. Many doctors will give barely a nod to the existence of medical school in their own past:

"Oh yeah, biochem and all that. I hated that stuff!"

"You'll learn it the right way as a resident anyway."

"Just gotta study and make it through!"

"Second year is the worst. After that, you start learning important things."

The implication is that medical school in and of itself is unimportant — a required step, of course, but not as important as the things that will come later. Thus, we all rationalize our way through it. How else can we make sense of the gratuitous amount of debt, the years of life invested and the massive delayed gratification?

However, while I am often guilty of it myself, I take issue with this mindset. I don't want to be an "in-between"; I don't want to see my education as a purely utilitarian object. I don't want these years of my life to be only a means to an end. The truth is, I am here right now: I am a thinking, living, breathing human, and my experience matters *right now*. My value isn't based in the fact that I will eventually be a doctor, that I will eventually make a big salary or that I will eventually know enough medicine to *really* help people. I have value because I am a person. That fact didn't change when I got accepted into medical school; it won't change if and when I finish medical school, either.

Basing our worth on our future is dangerous for two reasons. I've already mentioned the first: this mindset is close to dehumanizing medical students, the person who is here now. The second is the fact that none of us are guaranteed tomorrow. While it is tempting to believe our lives will have meaning in the future because of something we have earned, we can't depend on that. I was reminded of this while reading *When Breath Becomes Air*, the incredible memoir of Dr. Paul Kalanithi. He reminded me, in the most emotional and eloquent way imaginable, that life is beautiful and must be appreciated while we have it. Any one of us might reach the end of our training and find that we are unable to practice for one reason or another. This is a sobering truth, and while it is just as valid about other professions as it is for medical students, it is especially true for us because of the significant length and rigors of our medical education and training.

So, how do we fight against this mindset? How do we resist the temptation to see our education as a necessary evil and instead choose to see each day as the joy and blessing it is? I submit that there are many ways. We can, sometimes, take off the white coat and be with people not involved in medicine. We can protect time to spend with spouses, significant others, family and friends. We can take time for ourselves to rest, exercise, travel, read and employ other forms of self-care. We can resist the urge to make our only label "medical student."

And of course, we can write about our experiences. Communities like *in-Training* exist because medical students do have things to say. We write to process the things we see each day. We write to help other medical students know they are not alone. We write because we have stories to tell, stories that matter right now, because we matter right now. We write to remind each other and the world that we are more than just doctors-in-training: We are people with meaningful stories that just might be worth sharing.

Ria Pal grew up in San Luis Obispo, CA before developing her interests in neuroscience, community activism, and social justice at the University of Rochester in Rochester, NY over the course of her undergraduate and medical educations. She is profoundly grateful to the *in-Training* community for its many lessons in vulnerability and reflection. Currently, Ria is a child neurology resident at Stanford University. Her research interests include neurobiological consequences of trauma, bias in medical education, and effects of public art in childhood.

Andy grew up in Lake Geneva, WI. He is a former editor-in-chief of *in-Training* alongside Ria Pal and a current MD/PhD student at the Medical College of Wisconsin (Class of 2020), where he completed his PhD in Physiology. Andy received a BA in English at the University of Wisconsin - Madison.

www.ingramcontent.com/pod-product-compliance
Lightning Source LLC
Chambersburg PA
CBHW031810190326
41518CB00006B/266